Game Theory and the Law

Game Theory and the Law

Douglas G. Baird
Robert H. Gertner
Randal C. Picker

Harvard University Press
Cambridge, Massachusetts, and London, England   1994

Library of Congress Cataloging-in-Publication Data

Baird, Douglas G., 1953–
   Game theory and the law / Douglas G. Baird,
Robert H. Gertner, Randal C. Picker.
      p.   cm.
Includes bibliographical references and index.
ISBN 0-674-34119-8 (acid-free paper)
1. Law—Methodology.   2. Game theory.   I. Gertner,
Robert H.   II. Picker, Randal C.   III. Title.
K212.B35   1994
340'.1—dc20   94-8625
CIP

For
Bruce, Emily, and Lucy,
Leo,
and
Gretchen, Benjamin, and Adam

Contents

Preface

This book rests on the premise that game theory can offer insights to those who want to understand how laws affect the way people behave. Laws often matter in situations in which the behavior of one person turns on what that person expects others to do. Because strategic behavior is common, the formal tools that can help us understand it are important. We set out to accomplish two goals in writing this book. First, we wanted to introduce the formal tools of modern game theory to a wide audience using a number of classic legal problems ranging from tort and contract law to labor law, environmental regulations, and antitrust. These problems are familiar to those who are trained in the law and readily accessible to those who are not. Second, and as important, we wanted to show how modern game theory allows us to sharpen our intuitions and provides us with new ways of looking at familiar problems. In short, we have tried to write a book that offers those interested in law a new way of thinking about legal rules, and a book that shows those interested in game theory a fertile and largely unexplored domain in which its tools have many applications.

Much of the analysis in this book makes extensive use of concepts that have been developed only within the last decade, and we have not compromised on the rigor that these cutting-edge concepts demand. Nevertheless, we have been able to apply these concepts to the law without requiring the reader to know calculus, probability theory, or any other formal mathematical tools beyond simple algebra. Indeed, algebra is used only in a few places, and calculations are set out and explained in the endnotes. In this respect, this book stands apart from others that explore recent developments in game theory. We depend only on the reader's willingness to think through hard problems logically and carefully.

This book is the first to address in general terms the use of the formal tools of game theory and information economics in legal analysis, and many of the models and ideas it sets out are new. Nevertheless, legal scholars have worried about strategic behavior for a long time, and we have drawn on their insights. Moreover, there are many papers that take advantage of these tools in studying specific legal problems in subjects as diverse as procedure, contracts, conflicts of law, bankruptcy, taxation, and corporations. We have benefited from them as well. We have also relied heavily on work in law and economics and in game theory proper. The bibliographic notes at the end of each chapter give a sense of both the breadth of this literature and the connections between it and the ideas set out in each chapter. Given the extent of the literature, however, these notes are necessarily incomplete. The glossary defines the basic legal and economic terms that we use in the text.

We wish to thank the many who have helped us on this project. The research support of the Lynde and Harry Bradley Foundation, the John M. Olin Foundation, the Sarah Scaife Foundation, and the Graduate School of Business at the University of Chicago made this book possible. We received many useful comments in workshops at George Mason University, New York University, the University of Michigan, the University of Pennsylvania, and the University of Virginia.

In-Koo Cho, Cass Sunstein, and Alan Sykes offered valuable advice throughout the project. Patrick Bolton, Frank Easterbrook, Richard Epstein, David Friedman, and Mitchell Polinsky gave us the benefit of their close reading of several chapters. Anne Marie Harm prepared the figures. We are especially grateful to Ian Ayres, Jon Elster, Daniel Farber, Mark Fisher, Ted Frank, Jason Johnston, William Landes, Richard Posner, Mark Ramseyer, Robert Rasmussen, Elizabeth Rosenblatt, Eun Shin, David Skeel, and Lars Stole for their extensive written comments on the entire manuscript.

Game Theory and the Law

Understanding Strategic Behavior

Strategic behavior arises when two or more individuals interact and each individual's decision turns on what that individual expects the others to do. Legal scholars have long recognized the need to take account of strategic behavior. Too often, however, they have not taken advantage of the formal tools of game theory to analyze strategic behavior, other than to invoke a simple game such as the prisoner's dilemma as a metaphor for a collective action problem. This alone may not help us significantly. It may not matter much whether we call something a collective action problem or a multiperson prisoner's dilemma. This failure to make better use of game theory is unfortunate, given that modern game theory is sufficiently powerful to offer insights into how legal rules affect the way people behave. The challenge is one of applying its highly technical tools, many of which have been developed only in the last decade, to a new environment.

We begin in the first chapter with a problem—the tort rules governing an accident involving a motorist and a pedestrian—that is already well understood. Our ambition is to show how, even in the context of so simple a problem, a rigorous focus on strategic behavior advances our understanding of the legal rules. In the second part of the chapter, we review the best-known paradigms of game theory—the stag hunt, the prisoner's dilemma, and matching pennies—and show both how they may be applied to legal problems and what limits we encounter in doing this. In Chapter 2 we examine how parties interact with each other over time, using the problem of market preemption in antitrust law and several issues in debtor-creditor and contract law as vehicles for introducing additional game-theoretic tools.

In subsequent chapters we explore more difficult issues. Incomplete information is the central problem in game theory and the law. Shaping laws that give parties an incentive to act in a way that leaves everyone better off is a straightforward matter as long as all the parties and those who craft and enforce the legal rule possess enough information. Complications arise, however, when the necessary information is not known or, more commonly, is known, but not to all the parties or not to the court.

We focus in Chapter 3 on the simplest information problem. With respect to some kinds of information, legal rules have to confront a problem of unraveling—situations in which the ability of people to draw inferences from silence leads to the revelation of information. With respect to other kinds of information, we have to focus on the problems of signaling and screening—the ability of parties to draw inferences from the actions that other parties take. In both cases, we must take account of the ability of other parties and a court to acquire information. These distinctions should guide our understanding of such questions as legal rules governing the renegotiation of contracts. We develop a new model that shows how legal rules affect contracts that are written with the understanding that the parties may want to renegotiate contract terms at a later time.

We go on in Chapter 4 to analyze legal rules that must work in situations in which one party possesses information that cannot be communicated directly to other parties or to a court. Inferences about the information must be drawn from actions that the parties take. Advances in game theory now provide a rigorous way to explore how changes in legal rules affect the ability of parties to draw such inferences. Various legal regimes, such as plant closing laws and the Americans with Disabilities Act, can be put in this framework and subjected to scrutiny.

In Chapter 5 we examine how legal rules may affect parties who interact with each other over time. We illustrate this problem first by examining the Statute of Frauds, the contract law principle which requires that most contracts be evidenced by a writing. We show how a new way of interpreting this principle emerges once the problem is seen as a repeated game. We then examine a number of issues in antitrust law, including tacit collusion and predatory pricing. We use these problems to illustrate how legal rules affect the ability of parties to develop reputations that lead not only to anticompetitive conduct, but also to long-term cooperation.

In Chapter 6 we explore a number of different collective action problems. We show the dangers of treating interactions between parties as stand-alone games rather than as part of a larger game. Legal rules often address collective action problems when information is incomplete. We show how the mechanism design literature allows us to set out the limits of what legal rules can do in these situations. We also explore network externalities and herd behavior and the subtle ways in which the actions of one individual can impose costs on others.

How parties bargain with each other and the way in which they split the surplus from trade are problems that recur in legal analysis. Noncooperative bargaining theory provides a vehicle for understanding the dynamics of negotiations. In Chapter 7, we examine an important question of contract law involving the availability of specific performance when the loss a party suffers from breach is private information. We show how the problem can be seen as a bargaining game in which there is private, nonverifiable information, and we explain how this game can be solved. This model, like many others we develop in the book, is new. It suggests unappreciated strengths and weaknesses of a number of legal doctrines, including the automatic stay and the new value exception to the absolute priority rule in bankruptcy, and limits on the ability to hire permanent replacements in labor law.

In Chapter 8, we focus again on the dynamics of bargaining and, in particular, on the way in which legal rules governing bargaining must take account of the possibility that one or both of the parties may possess private information. We use a number of examples from the rules governing civil procedure to illustrate these problems. We also incorporate two-sided private information into a model of bifurcated trials and create a new model to explore rules governing discovery, a problem that by its nature turns on private information. In addition, we show that private information is likely to influence both the kinds of cases that are litigated and the inferences that can be drawn about the law from reported decisions.

Although we touch on many different legal problems in the course of the book and draw on much of the work that has been done in noncooperative game theory and information economics, particularly in the last decade, we also show that only a few basic paradigms are needed to capture the essential problems of strategic behavior that legal rules need to take into account. It is with the simplest paradigm—that of simultaneous decisionmaking under complete information—that we begin the first chapter.

Bibliographic Notes

The problem of strategic behavior. John von Neumann is generally recognized as the founder of modern game theory. Von Neumann (1928) and Von Neumann and Morgenstern (1944) are the two important early works. John Nash also made seminal contributions to game theory; see Nash (1950a) and Nash (1950b). Schelling (1960) is a classic nontechnical introduction to game theory, as well as one of the first works to enlarge significantly the scope of game-theoretic issues in the political and social sciences. A number of excellent introductions to game theory now exist. The best formal introductions include Fudenberg and Tirole (1991a), Kreps (1990b), and Myerson (1991). Fudenberg and Tirole (1991a) is encyclopedic and a useful reference for understanding many game-theoretic concepts. Kreps (1990b) connects the tools of game theory to the foundations of microeconomics. Hirshleifer and Riley (1992) is a good formal introduction to information economics.

Several excellent books on game theory aimed at nonspecialists have also been published in recent years. These include Binmore (1992), Dixit and Nalebuff (1991), Gibbons (1992), Kreps (1990a), and Rasmusen (1989). Both Binmore (1992) and Gibbons (1992) are formal introductions to game theory that are accessible to nonexperts. Gibbons (1992) focuses especially on applications of game theory to many standard economic problems, whereas Rasmusen (1989) focuses on the role of information. These books, however, require readers to be familiar with single-variable calculus and basic probability theory.

Dixit and Nalebuff (1991) is a nontechnical book that illustrates the general ideas of game theory using examples drawn from sports, business, and other familiar contexts. Kreps (1990a) discusses game theory and the conceptual underpinnings of its solution concepts.

Game theory and the law. The early applications of economic reasoning to the law were made with an eye toward understanding how legal rules affect behavior. The more prominent of these seminal works include Coase (1960) and Calabresi (1970). Posner (1992) offers a comprehensive examination of the economic analysis of law. First published in 1973, the book remains the starting place for analyzing many of the questions whose strategic components are the focus of this book. Farber and Frickey (1991) shows how public choice theory can illuminate legal analysis. See also Coase (1988) and Buchanan (1989) for more detail on Coase and the combination of legal issues and economic logic.

Legal scholarship that uses game theory in analyzing particular legal problems is large and growing. For example, Ayres (1990) provides a general discussion, and Jackson (1982) applies the prisoner's dilemma to bankruptcy law. Cooter, Marks, and Mnookin (1982) is one of the first studies to use an explicit game-theoretic model to examine what takes place before trial. Bebchuk (1984) and Bebchuk (1993) use information economics and game theory to examine rules of civil procedure. Mnookin and Kornhauser (1979) and Mnookin and Wilson (1989) examine strategic bargaining in the context of family law and bankruptcy respectively. Katz (1990b) uses game theory to analyze the problem of offer and acceptance in the law of contracts; Johnston (1990) uses it to explore contract default rules; and Gordon (1991) and Leebron (1991) use it to look at corporate law. Menell (1987) draws on the network externalities literature to analyze copyright protection for computer software. Brilmayer (1991) employs game theory to analyze problems in the conflicts of laws, as does Kramer (1990). Ellickson (1991) uses game theory to show how custom can work in much the same way as legal rules.

CHAPTER 1

Simultaneous Decisionmaking and the Normal Form Game

The Normal Form Game

The simplest strategic problem arises when two individuals interact with each other, and each must decide what to do without knowing what the other is doing. An accident involving a motorist and a pedestrian is such a case. The likelihood of an accident is determined by how carefully the motorist drives and how carefully the pedestrian crosses the street. Each must decide how much care to exercise without knowing how careful the other is. The behavior of the motorist and the pedestrian also depends on the legal regime. Legal scholars have long assumed that motorists will drive more carefully if they are liable for the injuries that the pedestrian suffers in the event of an accident. This observation alone, however, does not tell us how to shape the law of torts, and we must know more about how law affects these simple interactions if we are to understand its effects on more complicated ones.

Much of law and economics scholarship over the past several decades has focused on the intriguing claim that many legal regimes, including all those in Anglo-American tort law, induce the motorist and the pedestrian to act in a way that minimizes the total costs of the accident, costs that include the possibility of injury to the pedestrian as well as the expenses the motorist and the pedestrian incur when they take care to avoid the accident. To draw these conclusions, however, we need to make many assumptions. The pedestrian and the motorist, for example, must know what the legal rule is, and courts must be able to enforce it. Indeed, we cannot draw firm policy prescriptions or choose among possible tort regimes without subjecting all these assumptions to close scrutiny.

Many scholars have undertaken the task of exploring the various assumptions leading to the conclusion that these many different tort regimes are efficient. The debate about which rules work best when certain assumptions are relaxed now fills many volumes. We do not revisit this debate here. Rather, we begin by using the interaction between the motorist and the pedestrian to introduce one of the basic tools of game theory, the normal form game. We then show why many different legal regimes tend, under the same set of assumptions, to induce both parties to act in a way that is mutually beneficial. The model we develop in this chapter allows us to make clear exactly what it means to assume that individuals in the position of the motorist and the pedestrian are rational.

Game theory, like all economic modeling, works by simplifying a given social situation and stepping back from the many details that are irrelevant to the problem at hand. The test of a model is whether it can hone our intuition by illuminating the basic forces that are at work but not plainly visible when we look at an actual case in all its detail. The spirit of the enterprise is to write down the game with the fewest elements that captures the essence of the problem. The use of the word "game" is appropriate because one can reduce the basic elements of complicated social and economic interactions to forms that resemble parlor games.

Our goal in this chapter is to understand the common thread that unites different tort regimes. These regimes range from comparative negligence, in which liability is apportioned between parties according to their relative failure to exercise care, to strict liability with a defense of contributory negligence, in which the motorist is liable to compensate the pedestrian for any injuries unless that pedestrian acted carelessly. To discover what these different tort rules have in common, we can use a model in which the motorist and the pedestrian are each completely informed about everything, except what level of care the other will exercise. They know what it means to act carefully, and they know what consequences the law attaches to any combination of actions. Similarly, we may assume that courts can enforce any given legal regime and that they have all the information they need to do so. For many questions, of course, we cannot make so many simplifying assumptions, but, as we shall see, it is useful to do so here.

We model the interaction between the motorist and the pedestrian by using a traditional game theory model called a *normal form game*, sometimes referred to as the *strategic form* of a game. The normal form game consists of three elements:

1. The *players* in the game.

2. The *strategies* available to the players.

3. The *payoff* each player receives for each possible combination of strategies.

In the accident case, identifying the players is easy, at least if we avoid introducing such complications as whether one or both of the parties is insured. There are only two players, the motorist and the pedestrian.

The next step is to identify the strategies available to the players (or, to use a formal term, the *strategy space* of each of the players) by looking at the options that are open to them. Defining the strategy space is perhaps the most important decision in creating a model in game theory. The range of actions available to a motorist and a pedestrian is broad. Before we even reach the question of how fast the motorist chooses to drive, a choice that lies along a continuum, we face many others, such as the motorist's decision to buy a car, what kind of car to buy, and whether to go on a trip in the first place. Similarly, before choosing how carefully to cross the street, the pedestrian must first decide whether to go on a trip, whether to walk, which route to take, and where to cross the road.

How many of these possibilities are put into the model depends on what we want the model to do. A study of the forces at work in different tort regimes and the assumptions of rationality on which they depend requires only a strategy space in which the players must pick between two actions. In our model, therefore, each of the players faces only a binary choice—either to exercise *due care,* the amount of care that is socially optimal (by driving carefully or crossing the road carefully), or not to exercise care (by, for example, driving too fast or crossing the road without looking).

The last element of the normal form game is the payoff structure. We examine each possible combination of strategies and specify what happens to the pedestrian and the motorist in each case. Tort law is a regime of civil damages. It works by requiring one party to pay another damages under some conditions, but not under others. We can compare different legal regimes by taking games that are the same except for the division of the loss under each combination of strategies.

When the legal rule lets the losses lie with the pedestrian, the payoff to the pedestrian is the amount the pedestrian spends on care plus the expected cost of the accident, that is, the cost of an accident discounted by its likelihood. The likelihood of the accident turns, of course, on the particular combination of strategies that the players have adopted. The payoff to the motorist is simply the cost of exercising care. When the motorist is obliged to pay damages, the motorist's payoffs are reduced and the pedestrian's are correspondingly increased.

There are several different ways to represent payoffs. The basic idea we wish to convey is that the probability of an accident goes down as investment in care goes up, and the efficacy of one party's investment in care turns on whether the other party invests in care as well. The simplest way to do this is to posit dollar amounts for the costs of taking care and the costs of an injury that reflect these relationships. These amounts are intended to capture only the idea that the probability of an accident decreases as the parties put more effort into being careful, but that at some point the costs of taking additional care do not reduce the likelihood of an accident enough to justify them.

In the model that we build here, exercising care costs each player $10. An accident, when it occurs, inflicts a $100 injury on the pedestrian. We assume that the accident is certain to happen unless both players exercise care. (We could, of course, make a less extreme assumption about the need for both parties to take care, but again, the assumption we have made simplifies the problem without compromising our ability to study the effects of different legal regimes.) Finally, we need to make an assumption about the likelihood of an accident in the event that both the motorist and the pedestrian exercise care. We assume that, in this case, there is still a one-in-ten chance of an accident.

In a legal regime in which the motorist is never liable for the accident, if neither exercises care, the motorist enjoys a payoff of $0 and the pedestrian a payoff of −$100. If both exercise care, the motorist receives a payoff of −$10 and the pedestrian a payoff of −$20. (The pedestrian invests $10 in care and, assuming that the individual is risk neutral,[1] still faces $10 in expected accident costs, a one-in-ten chance of a $100 accident.) If the motorist exercises care and the pedestrian does not, the former receives a payoff of −$10 (the cost of taking care) and the latter a payoff of −$100 (the cost of the accident, which is certain to occur unless both take care). Finally, if the motorist does not take care and the pedestrian does, the motorist has a payoff of $0 and the pedestrian a payoff of −$110 (the pedestrian invests $10 in taking care and still suffers a $100 injury).

At this point, we have created a normal form game for the interactions between the motorist and the pedestrian in a world in which the pedestrian has no right to recover damages from the motorist in the event of an accident. An important step in modeling such an interaction is to take account of the information that the players possess. In the game between the motorist and the pedestrian, both know their own payoffs and those of the other player. They also know the strategies available to them and the strategies available to the other player. The only thing they do not know is the strategy the other player actually chooses (that is, they do not know whether the other player chooses to exercise care or not). This is a game of *complete but imperfect information*. If a player were unaware of something other than the strategy choice of another player, such as the payoffs the other player receives, it would be a game of *incomplete information*. It is also possible that both players know everything about the structure of the game and that one player can observe the strategy choice of the other player (as would be the case if the pedestrian could observe the motorist's care decision before determining how much care to exercise). In this situation, we would confront a game in which information is *complete and perfect*.

We can represent a normal form game involving two players who choose among a small number of different strategies with a *bimatrix*. In the bimatrix, each cell contains the payoffs to each player for any given combination of strategies. (This way of illustrating a normal form game is called a "bimatrix" because each cell has two numbers in it; in the ordinary matrix, each cell has only one.) Figure 1.1 illustrates our game using a bimatrix. By convention, the first payoff in each cell is the payoff to the row player, and the second payoff is that to the column player. In this figure, we assume arbitrarily that the pedestrian is the row player and the motorist the column player. Hence, the pedestrian's payoffs are given first in each cell.

The bimatrix is only one way of illustrating a normal form game. (A normal form game consists of players, strategies, and payoffs, regardless of how they are set out.) There are many normal form games that cannot easily be represented in this way. For example, we could create a model in which the motorist and the pedestrian choose from a continuum the amount of care to invest. Such a model would also be a normal

	Motorist	
	No Care	Due Care
Pedestrian No Care	-100, 0	-100, -10
Pedestrian Due Care	-110, 0	-20, -10

Figure 1.1 Regime of no liability. *Payoffs:* Pedestrian, Motorist.

form game, even though the number of strategies available to each player is infinite.

Now that we have reduced the interaction between the motorist and the pedestrian to a normal form game, we must "solve" it. We identify the strategies that the players are likely to adopt and then predict the likely course of play. Games are solved through the use of *solution concepts*, that is, general precepts about how rational parties are likely to choose strategies and about the characteristics of these strategies given the players' goals. Solving a game is the process of identifying which strategies the players are likely to adopt.

We must begin by making a fundamental assumption about how individuals make choices: Individuals are rational in the sense that they consistently prefer outcomes with higher payoffs to those with lower payoffs. We express payoffs in dollars, but this is not necessary. The basic assumption at the heart of this mode of analysis is not that individuals are self-interested profit-maximizers or care only about money, but rather that they act in a way that is sensible for them given their own tastes and predilections. This assumption may not always hold in an individual case, because people at times act in ways that are inconsistent and self-destructive. In general, however, people make the best decisions they can, given their beliefs about what others will do.

Once we assume that the behavior of individuals is rational in this sense, we can identify the strategy the motorist is likely to pick. In this game, taking care costs the motorist $10 and provides no benefit in return. The motorist always does better by not taking care than by taking care. We can predict the motorist's likely choice of strategy because there is a single strategy—taking no care—that, in the context of this model, is better for the motorist no matter what choice the pedestrian makes. Such a strategy is *strictly dominant*. A *dominant strategy* is a best choice for a player for every possible choice by the other player. One strategy is "dominated by" another strategy when it is never better than that strategy and is sometimes worse. When one strategy is always worse than another, it is "strictly dominated."[2]

This brings us to our first solution concept: *A player will choose a strictly dominant strategy whenever possible and will not choose any strategy that is strictly dominated by another.* This is the most compelling precept in all of game theory. Few would take issue with the idea that individuals are likely to choose a particular strategy when they can always do better in their own eyes by choosing that strategy than by choosing any other.

This solution concept by itself, however, tells us only what the motorist is likely to do in this model. We cannot use this concept to predict the pedestrian's behavior. Neither of the strategies available to the pedestrian is dominated by the other. It makes sense for the pedestrian not to take care when the motorist does not, but to take care when the motorist does. The pedestrian lacks a dominant strategy because either course of action could be better or worse than the other, depending upon what the motorist does.

To predict the pedestrian's behavior, we need to take the idea that players play dominant strategies one step further. Not only will a player likely adopt a strictly dominant strategy, but a player will predict that the other player will adopt such a strategy and will act accordingly. We can predict, in other words, that the pedestrian will choose a strategy based on the idea that the motorist will not choose a strategy that is strictly dominated by another. This bring us to our second solution concept, that of *iterated dominance: A player believes that other players will avoid strictly dominated strategies and acts on that assumption. Moreover, a player believes that other players similarly think that the first player will not play strictly dominated strategies and that they act on this belief. A player also acts on the belief that others assume that the first player believes that others will not play strictly dominated strategies, and so forth ad infinitum.*

This extension of the idea that dominated strategies are not played forces us to make a further assumption about the rationality of the players. We not only act rationally and do the best we can given our preferences, but we also believe that others act rationally as well and do the best they can given their preferences. This solution concept seems plausible if the number of iterations is small. After all, most people act rationally most of the time, and we can choose our own actions in anticipation that they will act this way. In addition, this solution concept relies on the assumptions we have made about the information that the players possess in a way that the strict dominance solution concept does not. With a strictly dominant strategy, a player does not need to know anything about the payoffs to the other player. Indeed, players do not need to know anything about their own payoffs, other than that the dominant strategy provides them with a higher payoff in every instance.

The greater reliance on both rationality and information is worth noting. When a legal rule provides both players with a dominant strategy, a player does not need to know any of the details of the legal rule or any of the effects of the legal rule on the other player. A player needs to know only that one course of action is optimal under all conditions.

As we move to solution concepts such as iterated dominance, we must assume more about what individuals know and more about how they believe others will act. The more we have to make such assumptions, the less certain we can be that our model will accurately predict the way individuals behave.

If we accept the iterated dominance solution concept, we can solve the game in Figure 1.1. The pedestrian will believe that the motorist will not exercise care because not exercising care is a dominant strategy. For this reason, the pedestrian will not exercise care either. Because this solution concept requires stronger assumptions about how individuals behave, however, we cannot predict the pedestrian's behavior as confidently as we can predict the motorist's.

We now face the question of whether we can draw any general conclusions from a model that reduces the problem to so few elements. The model is counterfactual in many respects. It assumes that, when there is no legal rule to shift liability to the motorist, the motorist incurs no costs and suffers no harm when an accident takes place. Most motorists are not indifferent to whether they run people over, quite apart from whether they are held liable. Given this and the many other similar assumptions, we want to know what general conclusions we should draw from the model.

Under our assumptions, once the motorist fails to exercise care, the accident will take place no matter what the pedestrian does. Although badly off by not taking care, the pedestrian is even worse off by taking care. We should not infer, however, that, as a general matter, pedestrians are likely to take no care when motorists fail to take optimal care. The model generated this result only because an accident was certain to occur unless the motorist exercised care. The interactions between investments in care on the part of the motorist and the pedestrian in another model might be quite different. The pedestrian might rationally take *more* rather than *less* care when the pedestrian believes that the motorist will take too little care. The motorist's failure to take care might make it sensible for the pedestrian to be even more vigilant.

The model, however, does contain a robust prediction. In a legal regime of no liability, the motorist will have too little incentive to take care. The motorist will take the optimal amount of care only if one makes implausible assumptions about the way people behave.[3] The motorist's strategy of taking the optimal amount of care—the amount of care that minimizes the total costs of accidents—is strictly dominated by a strategy of taking too little care. The amount of care the motorist takes will not be optimal. We would all be better off if we had a legal rule that induced both the motorist and the pedestrian to exercise due care. A legal regime in which the motorist is never liable contains a built-in bias that is likely to lead either to too many accidents or to unduly costly investments in care by the pedestrian.

This result in itself is hardly startling. To say that the strategy of taking reasonable care is dominated by another strategy of taking less than due care restates a familiar insight: individuals are more likely to be careless in a world in which people are not liable when they act carelessly. In such a world, people do not fully internalize the costs of their actions. The motorist enjoys all the benefits of driving fast but does not bear all the costs, namely, the danger of injuring a pedestrian. When we capture the problem of the pedestrian and the motorist in the form of a two-by-two bimatrix, however, not only are the incentives of the motorist made manifest, but, as we show in the next section, we can see how a change in the legal rules changes the incentives of the motorist and the pedestrian at the same time.

Using Different Games to Compare Legal Regimes

We can now use different variations of the game with the pedestrian and the motorist to compare legal regimes. We have the same players and the same strategies available to the players, but we change the payoffs, taking account of these different legal regimes. The payoff to the pedestrian under any strategy combination, in other words, now includes the expected value of the damage award the pedestrian receives, and the payoff to the motorist is lowered by the expected value of the damage award that must be paid to the pedestrian. The size of the expected damage award will be determined, of course, both by the liability rule for that particular combination of strategies and by the likelihood of an accident when the players choose those strategies.

Instead of a legal regime of no liability, let us go to the opposite extreme. The motorist is liable whenever there is an accident. This is a regime of pure strict liability. The motorist must pay for all of the pedestrian's injuries, regardless of whether either exercises care. This is the opposite of a regime in which there is no liability at all, in the sense that the motorist rather than the pedestrian bears the costs of the accident itself. This game is illustrated in Figure 1.2.

Exercising care still costs each player $10, and an accident is certain to happen (and cause $100 in damages) unless both exercise care. As before, there is a one-in-ten chance of an accident even if both exercise care. In this game, however, if neither exercises care, the motorist en-

	Motorist	
	No Care	Due Care
No Care	0, -100	0, -110
Pedestrian		
Due Care	-10, -100	-10, -20

Figure 1.2 Pure strict liability. *Payoffs:* Pedestrian, Motorist.

joys a payoff of −$100 and the pedestrian a payoff of $0. If both exercise care, the motorist receives a payoff of −$20 and the pedestrian a payoff of −$10. (The motorist invests $10 in care and still faces $10 in expected accident costs, a one-in-ten chance of a $100 accident.)

If the motorist exercises care and the pedestrian does not, the former receives a payoff of −$110 (the cost of taking care plus the cost of the accident) and the pedestrian a payoff of $0. (The pedestrian incurs no costs associated with taking care and is, by assumption, fully compensated for any injury and thus made whole in the event of an accident.) Finally, if the motorist does not take care and the pedestrian does, the motorist has a payoff of −$100 (the motorist takes no care but is liable for the cost of the accident) and the pedestrian a payoff of −$10 (the pedestrian invests $10 in taking care but recovers damages for the injuries arising from the accident).

Before solving this game, note the relationship between the games in Figure 1.1 and Figure 1.2: The sum of the payoffs in each cell in Figure 1.2 is the same as that in the corresponding cell in Figure 1.1. The change in the liability rule, however, reallocates the sum between the pedestrian and the motorist. We can capture the change in the legal rules by changing the payoffs, not by changing the strategies available to the parties themselves. Strategies represent those actions that are physically possible, whereas payoffs tell us the consequences of actions. Because the tort rules we are examining attach consequences to actions, they are reflected in the payoffs, not in the strategies.

We can discover the strategies that the players are likely to adopt by again using the solution concepts of strict dominance and iterated dominance. The pedestrian in this game (rather than the motorist) has a dominant strategy. The pedestrian will not exercise any care, regardless of what the motorist does. Under strict liability, the motorist must fully compensate the pedestrian for any injury. The pedestrian therefore bears none of the costs of the accident and ignores those costs when deciding whether to exercise care. The pedestrian takes no precautions because precautions are costly and bring no benefit. Exercising no care is a dominant strategy for the pedestrian. We can then invoke the concept of iterated dominance. The motorist recognizes that the pedestrian will act this way and therefore chooses not to exercise care as well. In this regime, the accident is certain to occur, the motorist is in all events liable, and the investment in care brings the motorist no benefit.

This model exposes a weakness of a regime of pure strict liability that parallels the one we saw in a regime of no liability. The pedestrian may have insufficient incentive to exercise care because the motorist bears the costs of the accident. A more elaborate model might provide for damages that do not fully compensate the pedestrian for the cost of the accident. Nevertheless, the general point of the model remains, even if these and other plausible complications are taken into account. The pedestrian will not fully consider the costs of the accident in deciding whether to take care, and hence may take too little care.

There are some accidents—such as airplane crashes—in which the victims have virtually no ability to take precautions. One might also favor a regime of pure strict liability in such cases if there were no way for a court to determine whether a victim took care. One would not, however, want such a regime in a situation in which the victim could take a number of readily visible steps to prevent an accident, and where it was therefore important to have a legal rule that provided the victim with an incentive to take care, just as one would not want a regime of no liability if the injurer needed an incentive to take care. Under a regime of either no liability or strict liability, it is likely to be in the self-interest of at least one of the parties to exercise less than due care.

We turn now to the legal regime of negligence plus contributory negligence, for a long time the prevailing principle of Anglo-American tort law. Under this regime, the pedestrian can recover damages only if the motorist is negligent *and* the pedestrian is not. This rule of law leads to the normal form game set out in Figure 1.3. As before, the legal rule does not change the strategies available to the players or the sum of the payoffs in each cell. All that changes is the allocation of the costs of the accident between the parties.

When we compare Figure 1.3 with Figure 1.1, we see that the two are identical except in the cell in which the pedestrian exercises due care and the motorist fails to do so. In this event, the pedestrian incurs a loss of only $10, the cost of taking care, and the motorist bears the $100 cost of the accident. The pedestrian continues to bear the cost of

	Motorist	
	No Care	Due Care
Pedestrian No Care	-100, 0	-100, -10
Pedestrian Due Care	-10, -100	-20, -10

Figure 1.3 Negligence with contributory negligence.
Payoffs: Pedestrian, Motorist.

the accident in the other three cases. The first two are cases in which the pedestrian fails to exercise care and the expected cost of the accident is $100. The third is that case in which both players spend $10 exercising care and the pedestrian also bears the $10 expected cost of the accident.

Unlike the game in Figure 1.1, this game is one in which the pedestrian has a dominant strategy. The pedestrian is always better off taking care. The motorist no longer has a dominant strategy. Whether the motorist is better off taking care depends on whether the pedestrian also takes care. If we accept the idea of iterated dominance, however, we can predict the strategy that the motorist will choose. The motorist recognizes that the pedestrian will exercise due care and therefore decides to take due care as well. Hence, under this legal regime, both pedestrian and motorist take due care. Our need to use iterated dominance to solve this game identifies a central assumption underlying this regime. To believe that a rule of negligence coupled with contributory negligence works, we must think that the motorist acts rationally and believes that the pedestrian acts rationally as well.

A comparison between the two models shows how this legal rule works. The only difference between Figure 1.1 and Figure 1.3, as mentioned, is in the cell representing the strategy combination in which the pedestrian exercises due care and the motorist does not. In Figure 1.1, the payoffs were −$110 and $0 to the pedestrian and the motorist respectively; in Figure 1.3, they are −$10 and −$100. This strategy combination is not the solution to either game, yet changing the payoffs associated with it completely alters the strategies that the parties adopt, and hence the expected play of the game. *The legal rule brings about changes even though it attaches consequences to actions that are never taken, either when the legal rule is in place or when it is not.* We shall return to this idea on a number of occasions. It will prove particularly important in those contexts in which some parties have information that others do not.

The legal regime in which there is strict liability, subject to a defense of contributory negligence, is set out in Figure 1.4. By injecting contributory negligence into the scheme of things, we make due care a dominant strategy for the pedestrian. Exercising due care results in a payoff to the pedestrian of −$10 instead of −$100. As long as the motorist believes that the pedestrian will play this strictly dominant strategy, the motorist will exercise due care as well, preferring a payoff of −$20 to one of −$100.

The difference between negligence coupled with contributory negligence and strict liability coupled with contributory negligence lies only in the consequences that follow when both players exercise care. In the negligence-based regime, the pedestrian bears the costs of an accident when both players exercise due care, whereas in the strict liability regime, the motorist does. The difference, however, does not affect the solution to the game, because exercising due care never costs a party more than $20, and failure to exercise due care under either regime when the other party does exercise due care leads to a loss of $100.

The comparison between regimes of negligence coupled with contributory negligence and strict liability coupled with contributory negligence in this model makes it easy to understand the well-known insight that both regimes give the two parties incentives to take care. It also unpacks the rationality assumptions that these rules need in order to work well even if we believe that everyone is well informed and that enforcement costs are low. We must assume not only that individuals behave rationally, but also that individuals expect others to behave rationally as well.

This way of looking at the problem reveals one of the important but subtle ways in which a legal rule works. A change in a legal rule can

	Motorist	
	No Care	Due Care
Pedestrian No Care	-100, 0	-100, -10
Pedestrian Due Care	-10, -100	-10, -20

Figure 1.4 Strict liability with contributory negligence.
Payoffs: Pedestrian, Motorist.

alter the behavior of both parties even by changing outcomes that are never seen under either the new or the old regime. Similarly, one can make a major alteration, such as changing the identity of the party who bears the loss when both exercise care, without affecting the incentives of either party.

Players choose the strategies that maximize their own payoffs. Hence, players compare their payoffs under a strategy relative to their own payoffs under other strategies. Changing the damages that one player must pay another alters the solution to the game only if the change makes a player's own payoff so much higher (or so much lower) that the player stands to do better (or worse) by choosing that strategy rather than another. A regime of negligence coupled with contributory negligence places the costs of an accident on the pedestrian when both parties exercise care, whereas a regime of strict liability coupled with contributory negligence places them on the motorist. These differences, however, are not large enough to affect the strategy choices of the players. One can change how losses are allocated when both parties exercise due care, as long as the consequences attached to exercising less than due care make that strategy choice less attractive.

The Nash Equilibrium

Regimes of negligence with contributory negligence or strict liability with contributory negligence have a sharp binary character. The pedestrian who falls just short of exercising due care receives nothing. Many have found this outcome normatively troubling and have advocated regimes of comparative negligence, in which both motorist and pedestrian shoulder some of the costs of an accident when both fail to exercise due care. A number of jurisdictions have adopted comparative negligence in their accident law. In this section, we examine comparative negligence regimes and, in the course of analyzing them, discuss another solution concept, the Nash equilibrium. A regime of comparative negligence may be harder to implement than the legal regimes it has replaced. Our focus, however, is again on how comparative negligence regimes differ from others as a matter of first principle. In this section, we examine comparative negligence with an eye to understanding how the incentives of the parties change once both bear some share of the liability when both fail to exercise due care.

The incentives that a comparative negligence regime imposes on parties depend on how liability is allocated when both parties fail to exercise due care. In some jurisdictions, the judge instructs the jury to allocate liability after considering "all the surrounding circumstances"[4] or "the nature of the conduct of each party and the extent to which each party's conduct caused or contributed to the plaintiff's injury."[5] Other sharing rules include one that looks at the amount of care that each party took relative to the total amount of care that should have been taken,[6] and another that reduces damages in the proportion which "the culpable conduct attributable to the claimant . . . bears to the culpable conduct which caused the damages."[7] The differences among the various sharing rules lead one to ask whether it matters which of these sharing rules one chooses. We can address this question by considering an extreme sharing rule, one in which the person who was the most careless bears a disproportionate share of the costs of the accident.

The players have a choice between exercising no care, some care, and due care. If both exercise no care or if both exercise some care, liability is split equally between them. If one exercises no care and the other exercises some care, however, the party who exercised no care bears a disproportionately large share and the party who exercised some care bears only a small portion of the costs of the accident. An accident imposes costs of $100 and is certain to happen unless both parties exercise due care.

Due care costs each player $3. Some care costs each party $1. When both exercise due care, the chances of an accident drop to one in fifty, and the expected costs of the accident to the pedestrian, the party who bears the costs when both exercise due care, is $2. If both parties exercise no care, they each bear the $50 loss. If both exercise some care, they each bear a $50 loss from the accident and $1 from the cost of the care they did exercise. If one exercises no care and the other exercises some care, the first bears $99 of the cost of the accident, while the other, more careful person bears only $1 in liability for the accident, plus the $1 cost of taking care that the player already incurred. The game is illustrated in Figure 1.5.

The incentives of the players are not so clear-cut under this comparative negligence regime as they were under the ones we examined earlier. Neither player has a strictly dominant strategy. In the comparative negligence regime set out in Figure 1.5, some care is usually worse than due care, but sometimes it is better than due care (in those cases in which the other player exercises no care). Neither due care nor some care dominates the other. Nevertheless, most people looking at this game have the intuition that both pedestrian and motorist will exercise due care.

Although each would be better off exercising only some care if the other exercised no care, neither the motorist nor the pedestrian expects

	Motorist		
	No Care	Some Care	Due Care
Pedestrian No Care	-50, -50	-99, -2	-100, -3
Some Care	-2, -99	-51, -51	-101, -3
Due Care	-3, -100	-3, -101	-5, -3

Figure 1.5 Comparative negligence (skewed sharing rule). *Payoffs:* Pedestrian, Motorist.

the other to exercise no care. The strategy combination in which one motorist exercises no care and the other exercises some care is not a likely course of play. While one player (the one who exercises some care) favors this strategy combination over all the rest, the other player (the one who exercises no care) does not, and therefore should be expected to choose a different strategy.

There are two formal ways of capturing this intuition and solving this game. The first idea is an application of the dominance ideas that we have already developed. No care on the part of the pedestrian and the motorist is dominated by due care. Because neither will play a strictly dominated strategy, we can reduce the game to a simple two-by-two game in which the only strategies on the part of each player are due care and some care. At this point, due care is a strictly dominant strategy for both motorist and pedestrian.

The second way we could solve this game is based on the following principle: *The combination of strategies that players are likely to choose is one in which no player could do better by choosing a different strategy given the strategy the other chooses. The strategy of each player must be a best response to the strategies of the other.* The solution concept based on this principle is known as a *Nash equilibrium.*[8] Introduced by John Nash in 1950, the Nash equilibrium has emerged as a central—probably *the* central—solution concept of game theory.

As applied to this game, this principle tells us that a strategy of some care on the part of either the pedestrian or the motorist and no care on the part of another is not likely to be the combination of strategies that the players adopt. Given that one of the players has adopted a policy of some care, the other player is better off using due care rather than no care.

The Nash equilibrium has loomed larger than dominance solvability because it can be usefully applied to more games of interest to economists. If the successive elimination of dominated strategies leads to a unique outcome, that outcome is also the unique Nash equilibrium of that game. A game that we cannot solve through the successive elimination of dominated strategies, however, often has a Nash equilibrium.

The ability of the Nash equilibrium concept to solve additional games makes it both more powerful and more controversial than solution concepts based on the idea that players do not choose dominated strategies. One can point to games in which the unique Nash equilibrium may not be the combination of strategies that players would in fact adopt.[9] Moreover, the Nash solution concept often does not identify a unique solution to a game. When there are multiple Nash equilibria, we may not be able to identify one of these as that which the players are likely to choose. Indeed, when there are multiple Nash equilibria, there is no guarantee that the outcome of the game is going to be a Nash equilibrium. Each player, for example, might adopt a strategy that is part of a different Nash equilibrium, and the combination of strategies might not be Nash. Nevertheless, the Nash solution concept is often useful in the context of a game such as this one and many others that we shall examine. We therefore focus on its formal definition more closely.

In a two-person game, a pair of strategies will form a Nash equilibrium when each player cannot do better given the strategy the other player has adopted. A Nash equilibrium, in other words, is a pair of strategies such that each is a best response to the other. To test whether a strategy combination forms a Nash equilibrium, let us call the strategy for the first player x^* and the strategy for the second player y^*. Now we need to ask whether, given that the second player will play y^*, the first player can do strictly better by switching to some strategy other than x^*. Similarly, we need to ask whether, given that the first player will play x^*, the second player can do strictly better by switching to some strategy other than y^*. If there is no better strategy for the first player than x^* in response to the second player's y^*, and if there is no better strategy for the second player than y^* in response to x^*, then this pair is a Nash equilibrium for the game.

Virtually all games of interest to us have at least one Nash equilibrium. More important, a strategy combination that is not a Nash equilibrium is unlikely to be the solution to the game. We can see this by assuming for a moment the opposite—that a particular strategy combination that is not Nash *is* the solution to the game. If such a combination is the solution to a game, both players should be able to identify this fact beforehand. If the strategy is not Nash, it follows, by definition, that one of the players is choosing a strategy that is not a best response given what the other player is doing. Put yourself in the position of the player whose strategy is not a best response. Why should you choose the strategy that is asserted to be part of the solution to the game? Given what the other player is supposed to do in this purported solution, you can do better. You are not acting rationally if you pick a strategy that does not maximize your own payoff.

If we return to the game that models a comparative negligence regime with a sharing rule that skews damages toward the party who was the most careless, we can see that only the strategy combination in which both players exercise due care is a Nash equilibrium. The strategy of due care for one player is always the best response when the other player exercises due care. If the players adopted any other combination of strategies, at least one of them would be choosing a strategy that was not a best response to the other, and that player could receive a higher payoff by switching to a different strategy.

Assume, for example, that the pedestrian exercised some care and the motorist exercised no care. The pedestrian has no incentive to exercise due care, given the motorist's strategy. The pedestrian prefers a payoff of −$2 to one of −$3. The motorist's strategy of no care, however, is not a best response to the pedestrian's strategy of some care. The motorist is better off exercising some care (and enjoying a payoff of −$51 rather than −$99) or exercising due care (and enjoying a payoff of −$3). The strategy combination in which both players exercise due care (and enjoy payoffs of −$5 and −$3 respectively) is the only combination in which neither player has an incentive to change strategy—or "deviate"—given the strategy of the other.

In this model, notwithstanding the extreme sharing rule, each player has the correct incentive. The model suggests that a comparative negligence regime is likely to induce both parties to exercise due care, independent of the particular sharing rule that a comparative negligence regime adopts. As others have shown,[10] even when the strategy space of each player is continuous, the only Nash equilibrium (and the only combination of strategies that survives the repeated elimination of dominated strategies) is the strategy combination in which both players exercise due care. As long as one accepts the Nash solution concept or the concept of iterated elimination of dominated strategies as a good prediction of the strategies that players will adopt, a comparative negligence rule gives players the right incentives in this simple environment, regardless of how the sharing rule is itself defined.

Civil Liability, Accident Law, and Strategic Behavior

The common law influences behavior by allowing injured individuals to bring actions for civil damages under specified circumstances. Such a legal regime stands in marked contrast to a regulatory regime that prescribes certain courses of conduct or subjects particular actions to criminal sanctions. A civil damages rule, seen through a game-theoretic lens, is simply a rule that reallocates the payoffs between players for each combination of strategies. The amount of wealth in each cell of the bimatrix remains the same, but it is distributed differently.

The power of a civil damages rule to affect the behavior of the players should not be underestimated. In games of complete but imperfect information, an infinite number of civil damages rules exists such that any outcome (including that which is the social optimum) is the only one that survives the iterated elimination of dominated strategies.[11] For this reason, it should come as no surprise that a number of different regimes (including the common law regime of negligence coupled with contributory negligence) provide players with the correct incentives as a matter of first principle. The common law is "efficient" in the limited sense that it gives the parties the correct incentives under a number of strong assumptions, but so do many other legal regimes. The interesting question is whether common threads unite these different rules of civil liability beyond the fact that they all induce parties to act efficiently under the same set of assumptions.

Regimes of negligence with contributory negligence, strict liability with contributory negligence, and comparative negligence all share three features:

1. The legal regimes are regimes of compensatory damages. Parties always bear their own cost of care, and the legal rules never require an injurer to pay more than is necessary to compensate the victim for the injury.

2. An injurer who exercises at least due care pays no damages whenever the victim does not exercise at least due care, and, in parallel fashion, a victim is fully compensated for any injuries suffered whenever the victim exercises at least due care and the injurer does not.

3. When both the injurer and the victim exercise at least due care, the costs of the accident are borne by one or the other or divided between them in some fixed proportion.

There are a number of other legal regimes that share these features as well. These include, for example, a legal regime in which the injurer is always liable when negligent and there is no defense of contributory negligence. They also include a regime of strict liability coupled with comparative negligence, in which the injurer is liable for the accident if both the injurer and the victim exercise care, but the losses are shared when both do not. There are other regimes that we do not see—such as those in which the losses are divided evenly between the parties when both exercise care—that also share these three features.

These three common characteristics are quite general. Regimes with radically different distributional consequences all share them. Nevertheless, they are themselves sufficient to ensure that both the injurer and the victim take due care. The proof of this proposition using the Nash equilibrium concept is the easiest to show, and we can set it out quickly.

Note first that excess care can never be part of a Nash equilibrium. If the other player exercises less than due care, a player avoids liability completely by playing due care. Excess care just creates costs and provides no additional benefit for this player. The costs of the accident have already been shifted to the first player. Alternatively, if the other player exercises due care, excess care cannot be a best response for a player, given how we have constructed our definition of due care. (If it were a best response, the due care-due care strategy combination could not be the social optimum.) We can therefore restrict our focus to strategies of due care or too little care.

The strategy combination in which both take care is a Nash equilibrium. Note that there are two generic allocations of liability when both players exercise due care: either one player bears the full costs of the accident, or the costs are shared. Consider the incentives of a player who does not bear the full costs of the accident. For this player, deviating from due care means bearing the full costs of the accident. For the deviation to be sensible, the gains from lowering the private cost of care must exceed, not only the extra expected costs of the accident, but also the additional fraction of those costs now borne by this player. This cannot happen, given how due care is defined.

A player is always better off exercising due care and bearing only a part of the expected costs of the accident than exercising less than due care and bearing all the costs of the accident. The increased costs of taking care are necessarily less than the reduction in the player's share of the accident costs. The reduction in the accident costs alone more than offsets the added costs of exercising due care rather than some lesser level of care. Any player who does not bear the full costs of the accident—namely, both players when costs are shared and the player bearing none of the costs when one player bears all of the costs—has due care as a best response when the other player exercises due care.

Consider finally a player bearing the full costs of the accident. This player cannot shift liability through the choice of strategy and thus just cares about minimizing the costs of the accident. Given that the other player is playing due care, the remaining social costs—all of which are borne by the first player—are minimized by selecting due care. Again, this follows from our definition of due care.

We must also ask whether any other combination of strategies could be a Nash equilibrium if a legal rule had these features. Consider any strategy combination in which one party is exercising due care and the other is not. In this event, the party who is not exercising due care bears all the costs of the accident itself and the costs of care which that party is taking. Thus, this party is better off deviating and exercising due care. Even if the party still bears all these costs, exercising due care leaves the party better off. The reduction in the expected accident costs necessarily offsets the costs of additional care. When one party exercises due care, the best response for the second party is always to exercise due care as well.

Consider finally the possibility that both players exercise less than due care. Either player could deviate, play due care, and incur only the costs of due care. We need to ask whether one party or another will have an incentive to play such a strategy. If so, the strategy of less than due care cannot be a best response for that player. In order for a strategy combination in which both players exercise less than due care to be a Nash equilibrium, two conditions must hold simultaneously. First, the injurer's share of the liability plus the injurer's cost of taking care must be less than the injurer's cost of taking due care. (The injurer's cost of taking due care is the relevant value for comparison. Given that the victim is exercising less than due care, the injurer can avoid liability completely by taking due care.) Second, the part of the injury that remains uncompensated plus the victim's cost of taking care must be less than the victim's cost of taking due care.

If both conditions hold at the same time, the costs to both the injurer and the victim together in this strategy combination are less than the costs to both of taking due care. In other words, the costs of the injurer's and the victim's taking care plus the costs of the accident—the social costs of the accident—must be less than the cost to the victim and the injurer of taking due care. This, however, cannot be true because, by definition, when the victim and the injurer exercise due care they minimize the total social costs of the accident. The costs of taking due care can never exceed the total social costs of an accident under any other combination of strategies. For this reason, one player would always prefer to exercise due care rather than less than due care in a strategy combination in which the other player was exercising less than due care as well.

We have now ruled out the possibility that any strategy combination in which one party does not exercise due care can be a Nash equilibrium. Therefore, the only Nash equilibrium in a game of complete but imperfect information in which the applicable legal regime satisfies these three conditions is the strategy combination in which both players exercise due care. Seen from this perspective, the various legal regimes that govern torts under Anglo-American rule are different variations on the same basic principle. Rules such as negligence, negligence coupled with contributory negligence, comparative negligence, or strict liability coupled with contributory negligence all share three very general attributes, which are themselves sufficient to bring about efficient outcomes in games of complete but imperfect information.

Under the strong assumptions we have been making, all the Anglo-American tort regimes induce parties to behave in a way that is socially optimal. They are all compensatory damage regimes in which a party never bears the costs of the accident if that party takes due care and the other does not. As long as a rule has these features, parties will have the right set of incentives. The rules have dramatically different distributional consequences, but these variations themselves do not give parties an incentive to behave differently.

Three general observations can be drawn from this examination of the common principle that links these different regimes. First, all these rules work in the same way and all depend on the same assumptions about the rationality of both injurers and victims. They require us to assume not only that individuals act rationally, but that they expect others to do so as well. Second, because these rules provide parties with the same set of incentives, choosing among the different rules requires us to examine all those things that are assumed away in this environment, such as whether a rule is likely to lead to more litigation or whether a court is more likely to make errors in enforcing a particular rule. We also cannot ignore the informational demands that the rules place on the parties. Parties do not need to know the particular content of the legal rule as long as they know that it is in their interest to exercise due care; all the rules, however, depend on at least one of the parties knowing what constitutes due care in any particular context.

Finally, this approach to the problem naturally leads to asking what other kinds of rules are possible. In the next section, we ask whether rules exist under which exercising due care is a strictly dominant strategy for both sides. If such a rule can be fashioned, we would not have to assume that parties expect each other to behave rationally. Each party would have the incentive to take care, regardless of what that party thought the other would do. We would not, of course, necessarily want to embrace such a rule if it existed, because it might come at too great a cost. Nevertheless, the first step in understanding how legal rules work is understanding what assumptions are essential to the enterprise.

Legal Rules and the Idea of Strict Dominance

In this section, we ask whether it is possible to state a rule of civil damages such that both players always find it in their interest to exercise due care, regardless of what each player expects the other to do. We begin by asking whether a regime can have this feature if it shares the same premise as those we have examined so far—legal regimes of compensatory damages in which a player who exercises due care never bears the costs of the injury if the other fails to exercise due care as well. We then ask whether other rules exist that are not built on this principle.

There is an intuitive way to describe the basic feature that we should see in a compensatory damages regime in which due care strictly dominates less than due care. Let us return to our example with the pedestrian and the motorist. The rule should ensure that the motorist and the pedestrian are always rewarded for the investments in care that they make, no matter what the other does. Hence, we want to make sure that both are better off for every dollar of additional care that they invest until they have invested in the optimal amount of care. In other words, their expected liability should go down by at least a dollar for each additional dollar they invest in taking care.

The game in Figure 1.5 proved difficult precisely because it did not have this feature. The pedestrian who invested only $1 in care was exposed to only $1 of liability when the motorist invested nothing. The pedestrian who invested in due care had to spend $2 more but would reduce the expected liability by only $1. When the motorist takes no care, the added costs to the pedestrian of taking due care instead of some care are greater than the benefits; hence, the pedestrian has no incentive to do so. Similarly, in regimes of negligence or strict liability coupled with contributory negligence, the motorist's investment in care brings no benefits to the motorist when the pedestrian exercises no care. A rule that ensures that both parties have an incentive to exercise due care no matter what the other does requires that the private benefits to a party from taking care always equal or exceed the private costs of taking care.

We can specify a sharing rule in a comparative negligence regime in which the costs that a party faces in taking care always correspond with the benefits that party receives in the way of reduced liability: A party who fails to exercise due care should bear the liability in proportion to the amount that party failed to spend on due care relative to the amount both parties fell short of exercising due care. Let us return to the example in Figure 1.5. Consider how this rule would allocate liability in the event that the pedestrian exercised some care (spending $1 instead of $3) and the motorist exercised no care (instead of spending $3). In this case, the pedestrian should have spent $2 more, and both parties together should have spent $5 more ($2 from the pedestrian and $3 from the motorist). Hence, the pedestrian should bear $2/5$, or 40 percent, of the liability for the accident. This rule does not allocate liability when both parties exercise due care. As the earlier discussion of negligence and strict liability suggested, the allocation of liability when both parties exercise due care does not affect the strategies that players adopt in a compensatory damage regime.

As stated, this sharing rule ensures only that exercising due care on the part of each party dominates all strategy combinations in which both parties exercise less than due care. The possibility that a player could exercise excessive care therefore must be taken into account. Once we do this, however, we discover that there is no compensatory damages rule of the type we have been discussing in which exercising due care is a strictly dominant strategy for both sides. Let us assume that the pedestrian is rational, but believes that the motorist is not and that the motorist will take excessive care. Excessive care reduces the likelihood of an accident. This may in turn lead a pedestrian who believes that the motorist will exercise excessive care to take less than due care. (Indeed, if the motorist actually did exercise excessive care, we might want the pedestrian to exercise less than due care.)

To modify our comparative negligence rule so that parties always find it in their interest to take due care no matter what they believe others will do, we need to modify this sharing rule to provide that a person also shoulders some of the liability when that individual exercises too much care. Such a rule is counterintuitive because the party who does this already bears the social costs of taking too much care.

The rule's justification lies in making it more likely that parties will take due care, not in the way it parcels out liability when one party acts contrary to self-interest. The obstacles that stand in the way of implementing this rule, including the difficulties of ascertaining due care in any case, are both obvious and substantial. In equilibrium, both parties would exercise due care. But the distribution of liability in combinations of strategies that are not part of the equilibrium are counterintuitive.

Consider the case in which the pedestrian crosses the street carelessly and is injured, even though the motorist exercised not merely due care, but excessive care. Under this rule, the pedestrian could sue the motorist and obtain a partial recovery of damages to the extent that the motorist was more careful than was socially optimal. Such an outcome seems wrong for two reasons. First, the motorist is punished even though the motorist bears all the costs of driving too carefully. It seems strange to force a party to pay damages to someone else when that party already bears the costs of departing from the social optimum. Second, given that the pedestrian did not exercise care, we may be better off if the motorist exercises excessive care. We can justify this allocation of damages only because we do not expect the motorist ever to exercise excessive care.

There is a simple rule that makes playing due care a strictly dominant strategy for both parties. The key is to relax the assumptions that damages be compensatory and that a party bears only the costs of taking care when that party exercises due care and the other does not. Consider the following regime:

1. Whenever both parties fail to exercise due care, and exercise instead too much or too little care, each party must bear some cost. (This requirement is usually trivial. If a party exercises any care or suffers some of the injury and has no right to recover damages from the other party, that cost is sufficient. The rule does require that the injurer pay some amount in damages in the event that the injurer exercises no care, even though this amount can be quite small.)

2. When one party exercises due care and the other does not, the latter must compensate the former for any injury suffered and must also reimburse the first party for the costs of taking care. (In other words, if the injurer fails to exercise due care, but the victim does, the injurer must compensate the victim not only for the injury, but also for the costs of care that the victim

incurred. Similarly, if the injurer exercises care and the victim does not, the victim must not only suffer the costs of the injury, but also compensate the injurer for the care taken.)

3. When both parties exercise due care, there is some rule allocating costs between them.

This rule is one in which due care is a strictly dominant strategy for both sides. Exercising due care strictly dominates any other strategy when the other player is not exercising due care. In all of these cases, exercising due care costs nothing and doing anything else costs something. Similarly, exercising due care dominates all other strategies when the other player exercises due care as well. Even if one party bears the full costs of the accident when exercising due care, that party is better off exercising due care than doing anything else. When the other player exercises due care, doing so must be the best response for the player. This player bears all the costs of the accident by exercising something other than due care. Hence, a player always has the incentive to exercise due care. Exercising due care minimizes the costs that the player faces, regardless of how much of the costs of the accident the player bears.

This rule has the same knife-edge characteristic that we see in rules incorporating negligence or contributory negligence. Such a rule can work only if we are confident that injurers, victims, and the courts can all identify the due care standard. Many other factors, such as the cost of legal error, need to be taken into account. It is, however, possible to create legal regimes in which parties must focus only on their own actions and do not have to take into account what others are likely to do. These regimes have the virtue of making fewer assumptions about individual rationality. They also should make us skeptical of relying on analyses of legal regimes that depend heavily on the same or similar assumptions and invoke the Nash equilibrium solution concept. If these assumptions hold, we do not even need the Nash equilibrium solution concept. In games of complete but imperfect information, the socially optimal outcome can be implemented with civil damages in strictly dominant strategies.

Collective Action Problems and the Two-by-Two Game

Unlike the game involving the motorist and the pedestrian, in which just two people were involved, many of the problems of strategic behavior facing a legal analyst are problems of collective action in which many individuals are involved. Nevertheless, these interactions often can also usefully be reduced to two-person games. Consider a problem that can arise in areas that are subject to flooding. At common law, flood waters are regarded as a "common enemy," and individual landowners have a right to build levees to keep flood waters off their land. This legal regime, however, creates a serious problem. Building a levee in one place increases the threat of flooding elsewhere. The response of individuals who are on the other side of the river or are upstream or downstream is to build new levees or increase the height of those that are already in place. In the end, investments in levees may not bring the landowners benefits commensurate with their costs relative to where they would be if no levees were built at all.[12]

In the game involving the motorist and the pedestrian, both players made their care decisions at the same moment in time. The game-theoretic problems involving simultaneous decisionmaking extend to a broader class of cases, however. They include any situation in which the players must act without knowing what the other player has done. Moreover, when enough people are involved so that negotiations between them are costly, the decision of each person may have little effect on the decisions of others. One may know what others do but have little ability to influence them. For such interactions, a simultaneous-move game may again be a useful model.

We can set out the essence of this problem with flooding by imagining that there are only two landowners, each of whom must independently decide whether to build a levee. We illustrate this game in Figure 1.6. If neither builds a levee, each will experience some flooding and suffer a loss of $4. A levee costs each landowner $2, but it eliminates the flooding problem only if the other does not build a levee. If one landowner builds a levee and the other does not, the landowner without a levee suffers a large flood and a $10 loss. If both build levees,

| | Landowner 2 | |
	Don't Build	Build
Don't Build	-4, -4	-10, -2
Build	-2, -10	-5, -5

Figure 1.6 Levee collection action game.
Payoffs: Landowner 1, Landowner 2.

they suffer $3 in flood damage. This amount is less than when neither builds a levee, but the landowners are worse off because the cost of the levee exceeds the amount saved from reducing the amount of flooding.

The two-by-two game that captures collective action problems like the one in Figure 1.6 is commonly called a *prisoner's dilemma*. The name comes from the story that was first told in the 1950s to illustrate the following strategic interaction: Two criminals are arrested. They both have committed a serious crime, but the district attorney cannot convict either of them for this crime without extracting at least one confession. The district attorney can, however, convict them both on a lesser offense without the cooperation of either. The district attorney tells each prisoner that if neither confesses, they will both be convicted of the lesser offense. Each will go to prison for two years. If, however, one of the prisoners confesses and the other does not, the former will go free and the latter will be tried for the serious crime and given the maximum penalty of ten years in prison. If both confess, the district attorney will prosecute them for the serious crime but will not ask for the maximum penalty. They will both go to prison for six years.

Each prisoner wants only to minimize time spent behind bars and has no other goal. Moreover, each is indifferent to how much time the other spends in prison. Finally, the two prisoners have no way of reaching an agreement with each other. Figure 1.7 reduces this story to a normal form game.

The games illustrated in Figures 1.6 and 1.7 have the same structure. Each landowner and each prisoner has a strictly dominant strategy—build a levee or confess. If the other landowner does not build a levee, the first can reduce flooding costs from $4 to $2 by building a levee. (There is no flood damage and the levee costs $2.) If the other landowner does build a levee, building a levee reduces losses from $10 to $5. (When the other landowner builds a levee and the first does not, the first landowner who does not build a levee incurs $10 in flood dam-

| | Prisoner 2 | |
	Silent	Confess
Silent	-2, -2	-10, 0
Confess	0, -10	-6, -6

Figure 1.7 Prisoner's dilemma. *Payoffs:* Prisoner 1, Prisoner 2.

age. When the first landowner does build a levee at a cost of $2, flood damage drops to $3 for a total cost of $5.) Either way, a landowner is better off building a levee. Similarly, a prisoner is much better off confessing than remaining silent if the other prisoner is going to confess. Six years in prison is preferable to ten years. A prisoner is even better off confessing if the other remains silent. By confessing, the prisoner can avoid prison altogether. No matter what the other prisoner does, a prisoner is better off confessing.

One would much rather not incur the cost of building a levee and suffer from a moderate flood than spend money on a levee and suffer from only slightly less flooding. Similarly, a prisoner would much rather spend two years in prison than six. These outcomes, however, are possible only when the players can reach a binding agreement. In both games each player has a strictly dominant strategy, and the strategy combination the players choose leaves them both worse off than they would be if they could cooperate with each other.

Collective action problems that fit the paradigm of the prisoner's dilemma present a possible case for legal intervention. For example, the government might have the expertise to build a system of levees that would minimize the costs of flooding to all the landowners as a group. As one court put it: "[T]he only adequate method of preventing this result was the unification of the individualistic and antagonistic efforts of the land owners on the opposite sides of the river into one comprehensive co-ordinating plan looking toward the flood control of the river in its entirety."[13]

This kind of problem is also generally known as a *tragedy of the commons*, named for the problem that arises when shepherds who share a common pasture overgraze it. Each shepherd does not incur all the costs of adding an additional sheep to the flock. Each additional sheep reduces the amount of grass available for the other sheep. The benefits to a single shepherd from grazing an additional sheep on the common pasture may be greater than the harm to the other sheep in that shepherd's flock, but smaller than the harm to all the sheep that graze there. Each shepherd enjoys all the benefits of grazing an additional sheep, but the harm to all the other sheep is borne by the shepherds as a group. Moreover, there are so many shepherds that the cost of reaching a consensual bargain among all of them is prohibitive. Hence, the shepherds collectively graze too many sheep on the common pasture.

The existence of transaction costs makes simultaneous decisionmaking an appropriate model for talking about this kind of problem. The model we used reduced the collective action problem to its barest elements. If we were interested in other questions (such as how the total payoffs in equilibrium change relative to the social optimum as the number of landowners changes), we would need to develop a more elaborate model in which there are many players.

The Problem of Multiple Nash Equilibria

We can illustrate the power of the two-person, two-by-two game by looking at another problem involving flooding and levees. In the previous example, we confronted landowners on opposite sides of the river. A different kind of problem can arise with landowners on the same side of the river. It may be in the interest of a landowner to build a levee and maintain it only if adjacent landowners build levees and maintain them as well. If any levee is improperly maintained, all landowners suffer damage in the event of a flood. A game that captures this problem is set out in Figure 1.8.

Maintaining a levee in this game costs $4, and a flood brings damages of $6. If both landowners maintain the levee, there is no flood, but both incur the $4 cost of maintaining the levee. If neither maintains the levee, they save money on maintenance but suffer flood damage of $6. If one maintains the levee and the other does not, the first suffers flood damage of $6 and incurs maintenance costs of $4, for a total loss of $10. The second suffers $6 from flood damage but incurs no maintenance costs.

Like many two-person, two-by-two games, this also fits within a well-known paradigm with a story attached to it. This game, known as the *stag hunt*, has two players who each have only two strategies: There are two hunters. Each must decide whether to hunt hare or stag. A hunter can catch a hare alone, but will catch a stag only if the other

| | Landowner 2 | |
	Maintain	Don't Maintain
Landowner 1 Maintain	-4, -4	-10, -6
Landowner 1 Don't Maintain	-6, -10	-6, -6

Figure 1.8 Levee coordination game.
Payoffs: Landowner 1, Landowner 2.

Hunter 2

		Stag	Hare
	Stag	10, 10	0, 8
Hunter 1			
	Hare	8, 0	8, 8

Figure 1.9   Stag hunt.   *Payoffs:* Hunter 1, Hunter 2.

hunter is also pursuing it. Sharing in half a stag, however, is better than catching a single hare.

The bimatrix takes the form shown in Figure 1.9. In this game, the strategy that each player adopts is good or bad depending on what the other does. If the first hunter were certain that the second would hunt stag, the first hunter would also decide to hunt stag. If the second hunter were going to hunt hare, however, the first hunter would hunt hare as well. The hunters' interests do not conflict. Each prefers to hunt stag, but only if the other does—and neither can be certain that the other will. Stag hunting will take place only if each is assured that the other will hunt stag.[14]

Solving either our second levee game or the stag hunt game introduces a new complication. The two landowners are each best off if both maintain the levee. The two hunters are best off if both hunt stag. The strategy combination in which both maintain the levee—or hunt stag—is a Nash equilibrium of this game. If the other landowner is maintaining the levee, the first landowner's best response is to maintain the levee as well. The first landowner should prefer a payoff of −$4 to a payoff of −$6. The second landowner is in a perfectly symmetrical position. We can engage in the same analysis for the stag hunt.

We cannot, however, be confident that both landowners will maintain the levee or that both hunters will hunt stag because games that have this structure have more than one Nash equilibrium. Consider the strategy combination in which neither landowner maintains the levee or in which both hunters hunt hare. If the other landowner is not maintaining the levee, the first landowner's best response is not to maintain the levee either. If one hunter is hunting hare, the other hunter's best response is to hunt hare as well. Once the other landowner is not going to maintain the levee, the first landowner is going to suffer from a flood whether or not the first landowner maintains the levee. The first landowner would rather suffer a loss of $6 than a loss of $10.

Similarly, once one hunter is not going to hunt stag, the other hunter will receive a payoff of $0 from hunting stag. The hunter would rather take the $8 payoff from hunting hare.

Making matters more complicated, games of this type have a third Nash equilibrium. So far, we have restricted our attention to "pure" strategy equilibria. A *pure strategy equilibrium* is a Nash equilibrium in which the equilibrium strategies are played with certainty, or with probability one. When the Nash equilibrium involves only strategies that are played with certainty, we have a pure strategy equilibrium. The alternative to a pure strategy equilibrium is a *mixed strategy equilibrium*, in which, in equilibrium, each player adopts a strategy that randomizes among a number of pure strategies.

An example of a mixed strategy would arise if one landowner randomly decided to maintain or not maintain the levee with equal probability. This particular mixed strategy, however, is not part of a Nash equilibrium. To see this, we need to discover the other landowner's best response to this strategy. The other landowner would calculate the expected payoffs from each of the pure strategies of maintaining and not maintaining the levee. A landowner always receives a payoff of −$6 when that landowner does not maintain the levee. We now must examine the payoff to this landowner from maintaining the levee when the first pursues this mixed strategy. By maintaining the levee (and incurring a cost of $4 in all cases and flood damage of $6 in half), this landowner receives an expected payoff of −$7.[15] Hence, a landowner's best response to this mixed strategy is not to maintain the levee.

We now know that, if the first landowner would maintain the levee half the time, the second landowner's best response would be to not maintain it at all. This strategy combination can be a Nash equilibrium only if maintaining the levee or not with equal likelihood is a best response to not maintaining it at all. It is not. When the other landowner does not maintain, this mixed strategy brings an expected cost of $8. (A loss of $6 half the time and a loss of $10 the other half.) The first landowner could do better by playing the strategy of not maintaining the levee with certainty and enjoy a payoff of −$6.

This may suggest how we find a combination of mixed strategies that is a Nash equilibrium. A player will be willing to randomize between two pure strategies only if that player is indifferent as to which of the strategies is played. A landowner plays the pure strategy of maintaining the levee if the payoff from it exceeds that from not maintaining the levee; the landowner plays the pure strategy of not maintaining it if the payoff from this strategy exceeds that from maintaining it. Hence, a landowner is willing to play a mixed strategy only if the payoffs from the two pure strategies are equivalent.

To understand how this works, return to Figure 1.8. We can see that, unless the first landowner is likely to maintain the levee, the second is better off not maintaining it. We want, however, to talk about this more precisely. Let p_1 be the first landowner's probability of maintaining the levee. There is a corresponding probability of not maintaining of $1 - p_1$. Let p_2 do the same for the second landowner. For the first landowner's given mixed strategy $(p_1, 1 - p_1)$, the second landowner will be indifferent between maintaining and not maintaining the levee if the expected payoffs from the two strategies are the same. The second landowner's expected payoff from not maintaining the levee is $-\$6$, independent of the first landowner's strategy. If the second landowner maintains the levee, the second landowner's payoff is $-\$4$ when the first landowner maintains it and $-\$10$ when the first landowner does not.

We determine the second landowner's expected payoff from maintaining the levee for any probability of maintaining or not maintaining it on the part of the first landowner by adding $-4 \times p_1$ and $-10 \times (1 - p_1)$. This amount is greater or less than the second landowner's $-\$6$ payoff from not maintaining, depending on the value of p_1. The second landowner's expected payoff from maintaining is equal to the payoff from not maintaining only when p_1, the first landowner's probability of maintaining the levee, has a certain value. This value is $2/3$.[16] The first landowner has to be twice as likely to maintain as not, or the second landowner will not be willing to give up a certain loss of $6 in exchange for the possibility of losing only $4, but risking a possible loss of $10.

When the first landowner adopts the mixed strategy of maintaining with $2/3$ probability, the second landowner is indifferent between maintaining and not maintaining. Given the first landowner's mixed strategy, any strategy the second landowner adopts, including any mixed strategy, is a best response. This game is symmetrical; hence, when the second landowner maintains with the levee $2/3$ probability, the first landowner is indifferent between maintaining and not maintaining or playing any mixed strategy. Any of these is again a best response to this mixed strategy.

When both landowners adopt the mixed strategy of maintaining the levee with $2/3$ probability, each is choosing a best response given the strategy of the other. The other landowner's decision to adopt this mixed strategy makes any strategy, including this mixed strategy, a best response. Therefore, this combination of mixed strategies is a Nash equilibrium. If one landowner were to adopt anything other than this mixed strategy in response to this mixed strategy on the part of the other, however, we could not have a Nash equilibrium. The first player's strategy would be a best response to the other player's mixed strategy, but the other player's mixed strategy would not be a best response to the strategy of the first.

Thus, the game in Figure 1.8 has three Nash equilibria, two in pure strategies and a third in mixed strategies. When a game has several Nash equilibria, it is not immediately self-evident how we should predict the strategies that the players will adopt. If there are ways to identify the one Nash equilibrium that individuals are likely to play and others that they are not, we may still be able to take advantage of the Nash equilibrium concept even when a game has multiple Nash equilibria. For this reason, much of the work in game theory over the last decade has focused on the question of whether we can isolate the kind of Nash equilibria that rational individuals are likely to play. We rely on these *refinements* of the Nash equilibrium concept when we examine a number of different legal rules in later chapters. At some point, however, we have to confront the limits of game theory. Although parties are likely to choose strategies that form a Nash equilibrium whenever a game has a predictable outcome, not all games have predictable outcomes.

One way of choosing among different Nash equilibria is to examine the different equilibria and ask whether any of them is especially prominent. Such a strategy combination is a *focal point.* It is also called a *Schelling point,* after Thomas Schelling, who examined this idea in an important early work on game theory.[17] The classical illustration of a focal point comes from experiments run several decades ago, in which a group of individuals were given the following thought experiment: You and another person must meet in New York on a particular day. You have no way of communicating with each other beforehand, however. You must therefore choose a time and location and hope against hope that the other person chooses the same time and spot.

This game has an infinite number of Nash equilibria. Every time and every location is a Nash equilibrium. Given that one player is in the middle of some block at some time during the day, the other player is better off being there at that time than waiting at any other place at any other time. Notwithstanding the infinite number of equilibria, however, the majority of those who participated in these experiments adopted the same strategy: they waited at noon at the information booth at Grand Central Station.

Those who engage in the same thought experiment today might not choose Grand Central Station. Grand Central Station no longer has the prominence it once had. In any given group, however, some focal point might exist. (Indeed, among game theorists familiar with the experiment, Grand Central Station may remain a focal point.) Returning to the game in Figure 1.8, one can argue that maintaining the levee is a focal point both because it is the outcome that brings the greatest benefit to the parties and because neither party is better off in any of the other Nash equilibria.

Experimental work on coordination games, however, suggests that players do not necessarily choose the Nash equilibrium that is in the individual interests of the parties and in their joint interest as well. Consider the game set out in Figure 1.10. There are two pure strategy Nash equilibria in this game—the strategy combination in which Player 1 plays up and Player 2 plays left, and that in which Player 1 plays middle and Player 2 plays center. Experiments suggest that individuals are overwhelmingly likely to choose the strategy combination of up and left, even though it leaves both players worse off than the combination middle and center.[18]

Players might adopt the Nash equilibrium that was in their joint interest if there were some possibility of preplay communication between the parties even if they had no way to reach a binding agreement. If two landowners each told the other that they were going to maintain their levees, each one might believe the other, because neither has anything to gain by persuading the other to adopt a strategy that is Nash and then deviating from it. If that other person actually adopts the

| | | Player 2 | | |
		Left	Center	Right
	Up	350, 350	350, 250	1000, 0
Player 1	Middle	250, 350	550, 550	0, 0
	Down	0, 1000	0, 0	600, 600

Figure 1.10 Coordination game experiment. *Payoffs:* Player 1, Player 2. *Source:* This game is taken from Cooper, DeJong, Forsythe, and Ross (1990).

strategy, the landowner's position cannot be improved by doing something else. Introducing these ideas of preplay communication, however, makes sense only if the parties can, in fact, communicate with each other. Moreover, such communication may not be effective if each player prefers a different Nash equilibrium and announces that preference to the other.

The danger that parties might not settle on the outcome that is in their joint interest even when it is a Nash equilibrium may provide a justification for a legal regime that changes the payoffs. For example, a legal rule that requires landowners to maintain levees once they build them would make the strategy of maintaining a levee a strictly dominant one for both landowners.[19] Landowners are made whole in the event that they maintain the levee and others do not, but they pay damages if they fail to maintain it when others do. Hence, maintaining the levee becomes a strictly dominant strategy for both landowners. We show the changes brought by a legal rule that requires landowners to maintain levees in Figure 1.11.

Parties who interact with each other can face other kinds of coordination problems as well. These can also be captured in two-by-two games. One such example is the problem of driving on one side of the road or the other. One type of driver might prefer the left-hand side of the road and the other the right-hand side, but each would rather drive on the less-favored side of the road if everyone else drove on that side as well. We see such problems of coordination in many places. In Chapter 6, we look at this problem in the context of the emergence of standards in an industry with several different firms. Legal rules may affect whether firms adopt a common standard and whether the one they adopt is the one that makes everyone better off.

The two-by-two game that captures this kind of problem is illustrated in Figure 1.12 and is typically called the *battle of the sexes*. It ac-

| | Landowner 2 | |
	Maintain	Don't Maintain
Maintain	-4, -4	-4, -12
Don't Maintain	-12, -4	-6, -6

Figure 1.11 Levee coordination game (with legal duty to maintain). *Payoffs:* Landowner 1, Landowner 2.

| | Spouse 2 | |
	Fight	Opera
Spouse 1 Fight	8, 4	3, 3
Opera	2, 2	4, 8

Figure 1.12 Battle of the sexes. *Payoffs:* Spouse 1, Spouse 2.

quired this name (obviously many years ago) because the story usually told to exemplify it was about a conflict between a couple who wanted to spend the evening together but had different preferences about whether to go to a fight or to an opera. Both would rather be with the other at the event they did not like rather than go alone to the event they preferred, but the first choice of both would be to go with their spouse to their favored event. Neither, however, is able to communicate with the other. Each must guess what the other will do.

It is a Nash equilibrium for both to go to the fight, for both to go to the opera, or for each to randomize between the two. In coordination games such as this, both players want to coordinate their actions, but each player wants a different outcome. To craft a legal rule that brings about cooperation in such cases, one must not only evaluate whether an outcome is efficient, but also weigh the competing interests of the players.

To this point, we have examined games where either there was a single Nash equilibrium in which parties adopt pure strategies, or there were multiple Nash equilibria. There are also games in which the only Nash equilibrium is one in which both players adopt mixed strategies. The prototypical game of this kind is *matching pennies*, illustrated in Figure 1.13.

| | Player 2 | |
	Heads	Tails
Player 1 Heads	1, -1	-1, 1
Tails	-1, 1	1, -1

Figure 1.13 Matching pennies. *Payoffs:* Player 1, Player 2.

Each player chooses heads or tails. The first player wins both pennies if both choose heads or both choose tails. The second player wins if one chooses heads and the other tails. This game is one in which there are no Nash equilibria in pure strategies. Given any combination of pure strategies, one player is always better off changing to the other. The only Nash equilibrium is one involving mixed strategies. Matching pennies is a classic *zero-sum game*, in which any gains to one player come at the expense of the other.

This paradigm most naturally applies to problems of law enforcement. Criminals have a powerful incentive to avoid acting in predictable ways. The same holds true for the police. Even if a law enforcement agency believes that a particular corner in a city is often used for drug deals, permanently placing a police officer on the corner will merely displace the deals elsewhere. If instead a random strategy is played, the police officer has a better chance of catching someone. Playing any strategy with certainty is likely to be foolish in the case of law enforcement—or indeed in any situation in which one party monitors another. The outcome of such games may well be a mixed-strategy equilibrium.

There is one other kind of strategic interaction worth noting. We explore it as a standard problem in negotiation. In negotiations one acquires an edge by being able to commit oneself to a particular strategy. If two players are splitting a dollar, the first player will acquire an enormous advantage if that player can make an offer that gives the other player almost nothing and at the same time make a credible commitment never to change the offer. The other player then has a choice between taking the little the first player offers immediately or rejecting it and never receiving anything. A problem can arise, however, if both players make such a commitment. A familiar problem in labor negotiations illustrates this point.

Parties to labor negotiations are not permitted to adopt a fixed policy of never deviating from the first offer they make.[20] To see why this legal rule may be sensible, consider the following situation. Each party to a labor negotiation can either commit in advance to making a single offer (and never being able to deviate from it) or to entering into a series of negotiations in which the parties engage in the ordinary process of exchanging offers and counteroffers.

We set out such a game in Figure 1.14. If the employer and the union each decide to bargain, there is no strike and each receives a payoff of $5. If one of them commits and the other does not, there is again no strike. Instead of an even division of the $10 surplus, however, the

Union

		Bargain	Commit
Employer	Bargain	5, 5	3, 7
	Commit	7, 3	2, 2

Figure 1.14 Collective bargaining. *Payoffs:* Employer, Union.

party who commits enjoys a payoff of $7 and the other enjoys a payoff of $3. If both commit themselves, there is a strike and each enjoys a payoff of only $2.

This is another game in which there are three Nash equilibria. Two are pure strategy equilibria in which one commits oneself to a fixed offer and the other is prepared to bargain. The third is a mixed strategy equilibrium in which, under these numbers, the employer and the union bargain a third of the time and commit themselves two-thirds of the time. Neither of the pure strategy equilibria seems to be a focal point because the combined payoffs are identical and the two players have strong and perfectly opposite preferences. Hence, this game may not have a predictable outcome. It is possible that the union and the employer will adopt either the mixed strategy equilibrium or some other combination of strategies in which both players sometimes commit themselves and a strike arises.

This two-by-two game differs from the battle of the sexes because the pure strategy Nash equilibria involve strategy combinations in which each player does the opposite of what the other player does. The general type of two-by-two game that captures this problem is known as *chicken*. The story behind this game comes again from an earlier time. Two teenagers drive cars headlong at each other. A driver gains stature when that driver drives headlong and the other swerves. Both drivers die, however, if neither swerves. Each player's highest payoff comes when that player drives head on and the other swerves; the second highest comes when that player swerves and the other player swerves as well; and the third highest comes when that player swerves and the other drives. The lowest payoff results when both drive. This game has multiple Nash equilibria, but unlike the stag hunt, the assurance game, or the battle of the sexes, the pure strategy equilibria are ones in which each player adopts a different action (that is, one swerves and the other drives).

We need to be cautious, however, about drawing any firm conclusions from our collective bargaining game. For example, it may not make sense to model the process as one in which the union and the employer decide independently of each other whether to commit themselves to an offer that cannot subsequently be changed. We may need to take account of the dynamic aspects of their interaction. The employer and the union may be able to negotiate with each other before either one makes an offer that cannot subsequently be reversed. To be useful, a model may have to incorporate these negotiations as well.

At this point, it is useful to acknowledge the strengths as well as the limits of using two-by-two normal form games to understand strategic behavior. Paradigmatic games such as chicken, the battle of the sexes, matching pennies, and the prisoner's dilemma can provide useful benchmarks. It may seem unduly limiting to begin with a game in which there are only two players and two strategies, but such simple games can capture the dynamics of many interactions. When we can use these games, the forces that are at work are readily apparent, and it is easy to understand the effects of different legal rules.

Nevertheless, one should use these paradigms with caution. First, one always wants to examine the strategic elements in a given situation and avoid being drawn too quickly to a well-known paradigm such as the prisoner's dilemma. Such paradigms can become Procrustean beds, and, by rushing to one or another too quickly, one may miss important parts of a problem. It is better to capture the problem in normal form and then look for the appropriate paradigm, rather than to shoehorn the problem into one at the outset. Taking advantage of a two-by-two game also requires an understanding of its limits. The prisoner's dilemma, for example, captures the basic feature of collective action and common pool problems, but a model with more elements will reveal details that the prisoner's dilemma does not. If one is interested in the dynamics of a particular collective action problem, the prisoner's dilemma may not be useful.

One must also guard against looking at interactions between players in isolation. A problem that may look like a prisoner's dilemma or some other simple two-by-two game may be part of a much larger game. One cannot assume that, once embedded in a larger game, the play of the smaller game will be the same. Moreover, many interactions between individuals are inherently dynamic. People deal with each other over time and make decisions in response to what the other does. Two-by-two games that model simultaneous decisionmaking are not useful vehicles for analyzing such problems.

Summary

The two-by-two bimatrix, and hence the familiar games that take this form, are well suited to analyzing the way legal rules affect the behavior of players when each must make decisions without knowing what the other will do. We have shown in this chapter how they provide a useful way to understand how different tort regimes operate and the strategic problems underlying specific issues in property law, labor law, and elsewhere.

We must compare two or more games in order to understand the effects of different legal regimes. The fewer the elements of each game, the easier it is to understand how they are different and how players in them might act differently. We should resist adding complications unless we are satisfied that they are necessary, for they tend to obscure the basic forces at work. The test of a model is not whether it is "realistic," but whether it sheds light on the problem at hand.

If a problem does not involve strategic behavior, we should not bring the tools of game theory to bear upon it. Similarly, when we encounter a problem of strategic behavior, we must be sure that the tool that we use is the appropriate one. Most important, we must first ensure that the problem to be analyzed dictates the tools that are used; second, we must use those tools that are best suited to the problem, rather than the ones that are the most accessible. Elegance and power are definite virtues of the two-by-two bimatrix, but these virtues may also lead to its being used in contexts for which it is unsuited, or for which other, more technically difficult tools are more suited. We begin to develop these in the next chapter.

Bibliographic Notes

The assumptions of game theory. As we emphasize in our discussion, game theory shares its basic premises with classical economics. For an elegant exposition of the basic principles of economics, see Becker (1971). Varian (1992) and Deaton and Muellbauer (1980) carefully explore different assumptions about preferences and choice. A good axiomatic discussion of decision theory is Kreps (1988). Kreps (1990b) introduces economics within a game-theoretic context. Elster (1986) provides an eloquent discussion of the limits of rationality and the theory of choice.

With the assumptions of game theory in hand, we can build a structure that does cast light on legal problems. None of this, however, is to suggest that these assumptions are trivial or unproblematic. For systematic criticism of the basic assumptions of economics, see Thaler (1991). Kahneman, Knetsch, and Thaler (1991) provides a general discussion of anomalies and the way in which the assumptions of economics and experiments appear to diverge. Kahneman, Knetsch, and Thaler (1990) looks specifically at the Coase theorem. For criticisms of the von Neumann-Morgenstern expected utility theory, see Hampton (1992).

Dominant strategies. A general discussion of dominance solvability and elimination of weakly dominated strategies may be found in Kreps (1990b, pp. 417–421). For a standard proof of the existence of a Nash equilibrium in the class of games we have considered in this chapter, see Friedman (1990, pp. 63–77). Several notes of caution should be made about our reliance on dominance arguments in this chapter. There is much debate over the significance of different solution concepts and their refinements. Although dominance arguments are arguably among the least controversial, they are not entirely free from controversy either. Nozick (1985) provides examples of games in which playing a dominant strategy leads to what might be considered an unreasonable result; see also Myerson (1991, pp. 192–195). In a similar vein, Cooper, DeJong, Forsythe, and Ross (1990) suggests that dominated strategies should not be entirely discounted in experimental situations.

Tort law and game theory. The torts literature is enormous, but for our purposes, three works stand out as benchmarks. Brown (1973) is generally credited with initiating the formal analysis of torts with his explicitly game-theoretic approach. Brown used elements of noncooperative game theory to explore different liability rules by invoking Nash solutions (although not under that name) rather than appealing to dominance solvability. Explicit use of game theory, however, has been limited. Landes and Posner (1987) and Shavell (1987) reveal the accumulated understanding of two decades' worth of economic analysis of torts. Neither text makes overt use of game-theoretic concepts. Landes and Posner makes no reference to formal solution concepts, and, although Shavell does use "equilibrium" (and even "Nash equilibrium"—once), he generally avoids such terms. Arguments made in these books, however, do rely implicitly upon the ideas of game theory. For example, Landes and Posner uses a dominance argument—without labeling it as such—in explaining why a defense of contributory negligence is unnecessary for the negligence rule to achieve the due care outcome. (See Landes and Posner (1987), pp. 75–76.) Landes and Posner (1981) provides a useful introduction to the economic theory of torts.

More recent works on torts have started to return to the explicit use of game theory. Examples include Orr (1991) and Chung (1992). Both argue for a comparative negligence standard on the basis of dominant strategy for both parties. The sharing rule for comparative negligence cases that we introduce in the text—and variations of it—have been mentioned or advocated in a variety of articles. Rea (1987) recognizes that it produces the correct marginal incentives; Orr (1991), Chung (1992), Sobelsohn (1985), and Schwartz (1978) also discuss the rule.

A large literature exists on the merits of different legal regimes. The case for comparative negligence is presented in Cooter and Ulen (1986), Calabresi and Hirschoff (1972), and Diamond (1974). There is also a large collection of works that examines tort standards from empirical and historical perspectives. See Curran (1992), Hammitt, Carroll, and Relles (1985), or White (1989).

The growth area in the recent torts literature has been the inclusion of friction in the standard accident models, either by relaxing the assumptions of perfect information or by making the legal system costly and/or unpredictable. On the cost of litigation, see Ordover (1978), Ordover (1981), Hylton (1990), and Hylton (1992). For an analysis of how errors by courts affect different legal regimes, see Hylton (1990), Polinsky (1987), Friedman (1992), and Rubinfeld (1987). Finally, Kornhauser and Revesz (1991) discusses the relative merits of negligence and strict liability in the context of an environmental tort.

The two-by-two game. The prisoner's dilemma was first discovered and recounted in its modern form in 1950 by scientists at RAND. As a matter of notation, we adopt the convention of referring to the game as the "prisoner's" dilemma, rather than the plural "prisoners' dilemma," because the emphasis, in our view, should be upon the individual and the choices which that individual faces. The dilemma is one that each prisoner confronts separately.

The history of the prisoner's dilemma and the social and political context in which it evolved are recounted in Poundstone (1992). Using an iterated prisoner's dilemma, Axelrod (1984b) examines issues of cooperation in the absence of legal institutions, while Axelrod (1984a) tackles that and other topics such as evolutionary stability and cooperation in WWI trenches. The general literature on the prisoner's dilemma is large. Some of the works that are most relevant given our perspective include Wiley (1988), Kreps, Milgrom, Roberts, and Wilson (1982), and Leinfellner (1986). Gibbons (1992) shows how the tragedy of the commons can be modeled as a normal form game with n-players and a continuous strategy space.

The stag hunt game can be traced back to an informal discussion of the problem by Jean-Jacques Rousseau. The strategic effects of Boulwareism are well known; see McMillan (1992). We discuss some additional issues in labor law in Chapters 3 and 7.

CHAPTER 2

Dynamic Interaction and the Extensive Form Game

The Extensive Form Game and Backwards Induction

In the last chapter we focused on the normal form game, which contains three elements—the players, the strategy space for each player, and the payoffs of each player for each combination of strategies. In this chapter, we examine the *extensive form game,* which models explicitly the actions that the players take, the sequence in which they take them, and the information they have when they take these actions. Because of its emphasis on actions, the sequence in which those actions are taken, and the information available to the players at each move, the extensive form game is often the appropriate way to model interactions between parties that take place over time.

We begin by using the extensive form game to model a simple debt contract. We go on to examine the problem of market preemption and strategic commitment in antitrust and then the problem of breach of warranty and mitigation in contract law. We use these examples once again to show how problems of strategic behavior can be captured in formal models and to introduce two new solution concepts—*backwards induction* and *subgame perfection*—that are central to the solution of games in which each player takes actions based on what the other player does. The incentives of the lender and the borrower in a straightforward debt contract illustrate the kind of problem we face.

Lender and Debtor consider the following bargain. Lender gives $10 to Debtor and Debtor promises to pay Lender back in a year's time. The trade is mutually beneficial. Lender has the capital and Debtor has the ability to use it productively. If all goes as planned, Debtor will use the loan to generate additional wealth and will be able to repay Lender in full, with interest. In Chapter 7, we shall examine the class of cases in which all does not go as planned and Debtor does not have enough to pay Lender back in full. At the moment, however, we assume that only Debtor's willingness to pay Lender back is in doubt.

After Lender makes the loan to Debtor, Debtor may not have sufficient incentive to pay the loan back. If Lender has no ability to call on the state for help in recovering the loan and no sanction to impose on Debtor, it may be in Debtor's self-interest to keep the money. Lender, recognizing that Debtor may not be inclined to repay the loan, might not be willing to make it in the first instance. Lender and Debtor need to create some mechanism that makes it in Debtor's best interest to pay the loan back when it is due.

Both Debtor and Lender are better off if Debtor's self-interest will lead to repayment of the loan. Unless Lender is confident that Debtor will repay the loan, Lender will not lend the money in the first place. Contract law is valuable because it makes it easier for parties to alter their incentives and make their promises *credible*. A promise is not valuable unless its beneficiary believes that it will be kept.

The extensive form game is a useful way to put a formal structure on this problem, and thus to gain a better understanding of it. An extensive form game contains the following elements:

1. The players in the game.
2. When each player can take an action.
3. What choices are available to a player when that player can act.
4. What each player knows about actions that have already been taken (by that player and others) when deciding to take an action.
5. The payoffs to each player that result from each possible combination of actions.

The most common way of illustrating an extensive form game is with an inverted tree diagram. Each possible point in the game is a *node*. By convention, the *initial node* is represented by a hollow circle, and subsequent nodes are shown as filled circles. The *branches* leading away from the node (typically shown as arrows) represent different actions available to that player. Each branch leads in turn to another node. If no branches lead away from that node, then no further actions can be taken should the game reach that point. At such a *terminal node*, there is a payoff for both players. Alternatively, the new node may be another *decision node*, at which a different player or the same player must again choose an action. Branches leading away from this node once again identify the different actions that are available to a player.

For the kinds of problems that are of interest to us, extensive form games are typically the best vehicles for examining interactions that take place over time, and normal form games are the best for problems of simultaneous decisionmaking. We should note, however, that any problem of strategic behavior can be set out in either form. For example, problems of simultaneous decisionmaking can be captured as extensive form games. The extensive game shows players moving in sequence. We can, however, treat a case in which players move simultaneously in the same way we treat problems in which one player moves after another, but moves without knowledge of what the other player has done.

Figure 2.1 is an extensive form representation of the normal form game that we saw in Figure 1.1. This is the game between the motorist and the pedestrian in which the pedestrian has no right to recover damages from the motorist in the event of an accident. In the extensive form of this game, the players are again the motorist and the pedestrian. The motorist makes the first move, choosing between two actions—either exercising no care or exercising due care. The pedestrian moves second, again deciding whether to take due care or not. The pedestrian is at either of two nodes, the node that arises after the motorist takes no care, or the one after the motorist takes due care. The pedestrian, however, does not know the course of the game up until this point.

We show that the pedestrian cannot distinguish between these two nodes by connecting them with a dotted line. By seeing which nodes are connected in this way, we can determine what each player knows about the actions of the other player at the time that player must move.

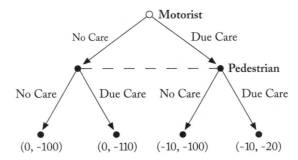

Figure 2.1 Regime of no liability (extensive form).
Payoffs: Motorist, Pedestrian.

When a player cannot distinguish between different nodes when that player must move, we describe all those nodes as being in the same *information set*. In the extensive form, a strategy for a player specifies the actions a player takes at every information set in the game.

The normal form of a game may have several different representations in the extensive form. For example, we could equally well have shown the pedestrian moving first and the motorist moving second. The solution to the game in either case, of course, is the same as in Chapter 1. This is a game in which the motorist has a dominant strategy (taking no care) and the pedestrian, recognizing this, will take no care as well. The strategy combination in which neither takes care is also the unique Nash equilibrium of the game.

An extensive form representation underscores different aspects of the game. In this case, the extensive form emphasizes that games of simultaneous decisionmaking are games of imperfect information. The players know everything about the game, except the moves that the other players make. The extensive form might have been the most natural one to use if we were looking at a tort problem that had parties acting in sequence. One such problem is the doctrine of last clear chance. This doctrine applies when the motorist can observe the care decision that the pedestrian has made. In such a case, the motorist has the duty to exercise care even if the pedestrian has not.

We can show this as an extensive form game by adding an additional move for the motorist after the pedestrian's move in Figure 2.1. However fast the motorist drove initially, the motorist could see whether the pedestrian was crossing the road carefully and then decide whether to slow down. The decision of whether to slow down would be branches added to what are shown as the terminal nodes in Figure 2.1. The nodes in which the motorist would decide whether to slow down would not be in the same information set (and hence would not be connected with a dotted line) because the motorist would know what action the pedestrian had taken when the time came to make this move.

Although the extensive form of this interaction may be easier to analyze, this game could also be represented as a normal form game. There is no longer a one-to-one correspondence between strategies and actions. A single strategy now consists of multiple actions. One such strategy, for example, would be to "exercise no care initially, slow down if the pedestrian exercises no care, do not slow down if the pedestrian exercises care."

Let us return now to the game involving an extension of credit and the problems that arise when players interact with each other over time. In this game, Lender and Debtor are the players. Lender and Debtor must each choose between one of two actions. (For Lender, the strategies are "Lend" and "Don't lend." For Debtor, the two strategies are "Pay back if Lender lends" and "Don't pay back if Lender lends.") The problems in this game arise because the parties move in sequence.

The extensive form represented in Figure 2.2a models the game between Lender and Debtor when Lender has no ability to sue Debtor or to use any other mechanism to control Debtor's postborrowing incentives. In this game, Debtor asks to borrow $100, promising to pay Lender $105 in a year's time. Lender moves first and decides whether to make the loan. If Lender does not make the loan, both parties receive a payoff of $0. If Lender does make the loan, the money can be used in Debtor's business and Debtor can earn $110 over the course of the year. At this point, Debtor must decide whether to pay Lender back. If Debtor repays Lender, both will enjoy a payoff of $5. (Lender will enjoy $5 in interest on the loan and Debtor will enjoy the profits that remain.) If Lender makes the loan and Debtor defaults, however, Lender loses $100 and Debtor gains $110.

It is in the joint interest of the parties that the loan be made, because when the loan is made, the parties enjoy a joint payoff of $10 rather than $0. We can see that, however, if each party acts out of self-interest, the loan will not be made and a mutually beneficial trade will not take place. This conclusion is immediately self-evident in a case such as this, but the process that we use here, backwards induction, can be used in more complicated cases as well.

We start by focusing on the reasoning process of Lender, the player who makes the first move. Lender will determine what Debtor will do when given the move and then reason backwards before deciding

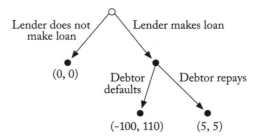

Figure 2.2a Lending without legal enforcement of debt contracts. *Payoffs:* Lender, Debtor.

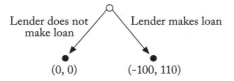

Figure 2.2b Lending without legal enforcement of debt contracts (inducted). *Payoffs:* Lender, Debtor.

whether to make the loan. Hence, we look, as Lender would, at the choice that Debtor faces in the event that the loan is made. Because, by assumption, Lender has already made the loan, Debtor must simply weigh the two possible payoffs, $110 and $5. The first dominates the second. Because $110 is more than $5, Debtor will choose not to repay the loan if given the choice.

Now that we know how Debtor will move if given the chance, we can truncate the decision tree and eliminate the strategy that we know Debtor will not adopt. The truncated version appears in Figure 2.2b. Once the game tree is truncated in this way, Lender can readily see the consequences of making the loan. Because $0 is better than a loss of $100, Lender chooses not to make the loan. Because this outcome is not in the interest of either Lender or Debtor before the fact, they will seek to transform this game into one that has a different solution. A contract can be understood as a mechanism designed to transform this game into one in which the trade goes forward.

A legal rule that allows Lender to call upon the state to enforce a loan in the event of default and make Debtor pay its attorney's fees may make contracting between the parties possible. The legal rule not only prevents Debtor from refusing to repay the loan, but also exposes Debtor to the costs of litigation. Debtor might not contest the lawsuit and incur no costs at all, but let us assume that this debt contract, like most, requires that Debtor reimburse Lender for the latter's litigation costs, which are $10. The legal rule transforms the game between Lender and Debtor into the game in Figure 2.3.

We can again use backwards induction and look at the decision facing the last player to move. At the time that Debtor must decide whether to repay the loan, Debtor is better off repaying than defaulting. Debtor has no ability to keep the money, and any effort to keep the money will expose Debtor to $10 in litigation costs, which more than offsets the $5 profit Debtor would make if the loan were repaid. Because Debtor will be better off repaying the loan, Lender is better off

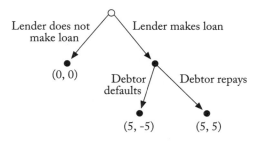

Figure 2.3 Lending with legal enforcement of debt contracts. *Payoffs:* Lender, Debtor.

making the loan in the first instance. By changing the payoff to Debtor in the event of default from a gain of $110 to a loss of $5, the game becomes one in which Debtor is led by self-interest to repay the loan, and, more important, Lender is led by self-interest to make the loan in the first place.

This example illustrates why it might be in the interests of both parties to allow one to call upon the state to enforce a promise that the other made. It does not, however, tell us that contract law should apply to all agreements; nor does it say anything about the content of contract law. Indeed, it suggests that contract law should be largely an empty vessel. Parties need to be able to ensure that a person who is the victim of a breach has the right to obtain a legally enforceable order that requires the other to pay damages. This ability gives the parties the power at the start of their interaction to alter payoffs arising from particular strategy combinations and to ensure that both parties adopt those strategies that work to their mutual benefit.

Every trade is different, and the constraints that work on parties outside the law are different as well. As long as the agreement between two parties imposes no costs on anyone else, contract law may work best if it simply gives the players the license to perturb the payoffs in a way that advances their mutual interest. Any other method of enforcing contracts may not change the payoffs in a way that guarantees that each party looks out for the interest of the other.

A law that gives Lender the ability to call upon the state to enforce its claim provides parties with a way of transforming a game with a suboptimal equilibrium into another game with an optimal equilibrium. It is important, of course, not to overstate the point, and there are many qualifications that we should make. Most obviously, to say that legally enforceable contracts facilitate mutually beneficial trade is not to say that the existence of such trade depends on it. Trade can exist and indeed flourish in the absence of legally enforceable contracts. As we shall see in Chapter 5, mechanisms such as reputation can bring about long-term cooperation even if there is no enforceable contract. The prospect of losing future deals both with another party and with people who know that party may be sufficient to ensure that each party performs.

We need to be cautious about the virtues of legal enforceability for a second reason. Making a contract legally enforceable is rarely as easy as the game in Figure 2.3 might suggest. In that game, Lender's money was returned with interest, regardless of whether Debtor defaulted, because we assumed that Lender's collection costs were recoverable in full. When Lender cannot recover such collection costs, the situation becomes more complicated. When the time comes for repayment, Debtor may be able to offer to pay less than what is owed in return for freeing Lender from the costs of litigation. The possibility that Lender may subsequently find it most beneficial to renegotiate the original contract is one of the most important issues in contract law. We shall explore the problem of renegotiation in Chapter 4. At this point, however, we wish to continue examining the extensive form game and, in particular, the solution concept of backwards induction. We do this in the next section by turning to the problem of market preemption and strategic commitment in antitrust.

A Dynamic Model of Preemption and Strategic Commitment

The case of *Federal Trade Commission v. E. I. DuPont de Nemours and Co.* focused on whether DuPont had acted improperly in the market for titanium dioxide,[1] a whitener used mainly in paints and plastics. By the 1970s, DuPont had developed a proprietary manufacturing process that made it the undisputed cost leader in the industry. The company then embarked on a plan to increase its production capacity substantially. Indeed, it planned to increase production by an amount sufficient to enable the company to capture much of the growth in the domestic demand for titanium dioxide through the 1980s. The government brought a complaint against DuPont before the Federal Trade Commission (FTC) in which it asserted, among other things, that the adoption and implementation of these expansion plans were themselves "unfair methods of competition" within the meaning of §5 of the Federal Trade Commission Act.

The threshold question we face is determining exactly how expanding capacity itself could be an "unfair method of competition." Instead of looking at DuPont and the worldwide market for titanium dioxide, we shall look at a problem involving a firm with market power in a small region. Assume that in a small town there is only one cement plant, which we shall call "Incumbent." Trucking in cement from a more distant town is not practicable. For this reason, Incumbent can charge higher prices than cement plants in towns that face competition. Incumbent is very successful and earns profits of $25 per month. Incumbent has the option of expanding its plant by purchasing land adjacent to it.

Another firm, which we shall call "Entrant," owns land in the town. It must decide between building a competing cement plant or a completely unrelated manufacturing plant that will have no effect on the market for cement. We want to know whether Incumbent will expand its cement plant and whether Entrant will build a second cement plant or a manufacturing plant. The benefits each enjoys from either course of action, of course, depend on what the other does.

We shall attach payoffs to the four possible combinations of decisions by Entrant and Incumbent respectively: (manufacturing, expand); (manufacturing, maintain current size); (cement, expand); and (cement, maintain current size). Entrant earns $10 if it opens the manufacturing plant. This amount remains constant no matter what Incumbent does. The amount that Entrant earns from opening a second cement plant in town, however, does turn on whether Incumbent expands. Entrant earns $15 if Incumbent maintains its old size, but it earns only $5 if Incumbent expands. The profits are much smaller if Incumbent expands because Entrant will have to cut its prices to compete.

Incumbent's payoffs depend on Entrant's decision. If Entrant does not build a competing plant, Incumbent earns only $25 if it maintains its current size, but $30 if it expands. If Entrant opens a cement business, Incumbent is better off at its current size, where it earns $10, rather than expanding, in which case it earns only $5. In the latter case, not only does Incumbent have a smaller share of the market, but it also incurs the costs of maintaining a larger plant.

So far we have said nothing about the timing of the decisions. Assume that Incumbent cannot make its choice until after Entrant, perhaps because it will take several months to acquire the necessary permits. We can now represent the game by the extensive form in Figure 2.4.

It is easy to use backwards induction to solve this game. When In-

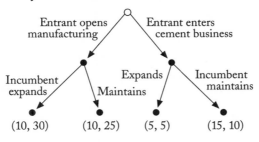

Figure 2.4 Entrant moves first. *Payoffs:* Entrant, Incumbent.

cumbent is at the left node, it prefers the strategy of expanding. The left-hand node is the one at which Entrant has already decided to go into the manufacturing business. By contrast, Incumbent maintains its current size at the right node, the one at which Entrant opens a second cement plant. Anticipating these decisions, Entrant has a choice between the terminal nodes at the extreme left and right in Figure 2.4. It prefers the one on the extreme right, which gives it a payoff of $15 rather than $10. Thus, the equilibrium is for Entrant to open a cement business and for Incumbent to maintain its current size.

The outcome (cement, maintain) does not maximize the two firms' joint profits nor is it the one that Incumbent prefers. Incumbent wants Entrant to believe that it will expand even after Entrant chooses to compete in the cement business. If Entrant believed that Incumbent would in fact expand even after it opened a cement plant, Entrant would be better off building the manufacturing plant (and thereby enjoying a payoff of $10 rather than $5). In this case, Incumbent could expand and enjoy profits of $30 rather than the profits of $25 it would receive if it maintained its current size.

Entrant, however, knows that Incumbent will not expand once it has opened its cement plant. Moreover, Incumbent cannot threaten to expand regardless of what Entrant does. A threat will be believed only if it is credible. Incumbent's threat to expand after Entrant opens a second cement plant will not be believed because, at that point, Incumbent makes itself worse off by expanding.

In the game set out in Figure 2.4, Incumbent has no way of convincing Entrant that it will expand even if Entrant opens a second cement plant. In many cases, however, Incumbent does in fact have strategies that allow it to commit itself to expanding before Entrant decides whether to enter the manufacturing or the cement business. To put the point another way, the sequence of moves determines the outcome of

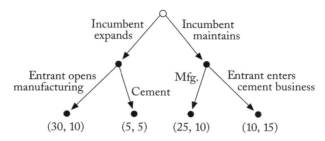

Figure 2.5 Incumbent moves first. *Payoffs:* Incumbent, Entrant.

this game. Incumbent would be better off moving first and therefore will be willing to spend resources to do so.

We shall soon explore how Incumbent might accelerate its decision, but for now let us just assume that Incumbent moves before Entrant. The extensive form of this game is depicted in Figure 2.5. By convention, the payoffs are written in the order in which the players move. Hence, in this game, the payoffs to Incumbent are set out first.

We can use backwards induction to solve this game. If Incumbent chooses to expand its plant, Entrant will enter the manufacturing business (10 > 5), whereas if Incumbent chooses to keep its cement plant at its current size, Entrant will open a competing cement plant (15 > 10). Incumbent will therefore choose to expand because it predicts that Entrant will open a manufacturing plant in response. This course gives Incumbent $30. By contrast, if Incumbent decided to maintain its plant at its current size, it would induce Entrant to open a new cement plant, and it would then enjoy profits of only $10.

As the contrast between these two games suggests, a game's likely course of play is often determined by which party moves first. (In these models, the player who moves first enjoys the advantage, although in some strategic settings moving last is advantageous.) Incumbent therefore wants to commit itself to expanding before Entrant decides whether to enter the cement business.

There are many ways of making such a commitment. Although the adjacent land may not become available until much later, Incumbent can contract to buy it now. It can sign contracts for equipment for the new plant, advertise the expansion to customers, and begin to hire and train more employees. These strategies create commitment in two ways. First, some of the policies move the timing forward so that Incumbent's decision precedes Entrant's. Expansion prior to when the capacity is needed is known as *market preemption*. Second, some of these policies change the payoffs by reducing the relative costs of expansion. Two ingredients are necessary for this second tactic to be successful: Incumbent must bear some of the costs of expansion before Entrant's decision, and these costs must in part be sunk, so that Incumbent will not recover them even if Entrant were to go into the cement business. (As long as Incumbent can get its money back, its threat to expand is not credible and Entrant will find it in its interest to open a second cement plant.)

Let us say that Incumbent can invest money that would otherwise bring a return of $10 per month on specialized equipment needed only if the plant expands. Incumbent's profits are now $10 lower if it does not expand. Incumbent has already spent the money that it would not ordinarily spend until after it expanded the plant. Once this investment is sunk, it has no value unless Incumbent expands. The new payoffs are set out in Figure 2.6.

Expanding the plant is a dominant strategy for Incumbent. Entrant anticipates expansion by Incumbent and therefore chooses the manufacturing business, making the outcome (manufacturing, expand). Incumbent spends the same amount on expansion in this game as in the ones set out in Figures 2.4 and 2.5, except that it spends some of the money earlier. Incumbent is better off making a preemptive investment in expansion, even if it costs more, as long as the investment is large enough to make expansion a dominant strategy and the payoff from expanding exceeds $10. Incumbent is therefore willing to spend up to $20 more in order to keep Entrant from going into the cement business. This $20 represents the difference between the profits in the equilibrium in which Incumbent could move first and the equilibrium in which Entrant could move first.

Some types of commitment do not waste resources. If Incumbent

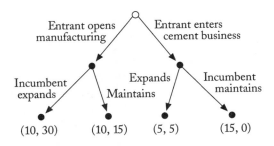

Figure 2.6 Incumbent sinks investment in expansion. *Payoffs:* Entrant, Incumbent.

merely signed a contract to lease the space, for example, its costs would be sunk, but someone else can use the space before Incumbent's lease begins to run. On the other hand, if Incumbent bought equipment for the larger plant and kept it idle in storage for a time, resources would be wasted. In principle, everyone would be better off if somehow Incumbent did not make such expenditures. To be sure, the money that is being spent belongs to Incumbent, but the private gain to Incumbent exceeds the social gain from this investment. Incumbent is willing to make investments that bring a deadweight social loss as long as the loss is less than the private benefit Incumbent enjoys from keeping Entrant out of the market.

In the model, Incumbent and Entrant are both better off if one expands and the other enters a different business. The model, however, does not show us how this course of action affects consumers. It is possible, for example, that people in the town are better served by one large cement plant than by two smaller ones. The profits might be larger with the expanded plant because of economies of scale, not because of monopoly power. The prices may not be higher if there is only one cement plant rather than two, given the amount of cement that people in the town buy. Even if Incumbent expends resources ensuring this outcome, consumers as a group may not be worse off.

This model shows why DuPont had reason to commit itself to building capacity beyond its current needs. Once it had built the facilities needed to meet projected increases in demand for titanium dioxide, its costs were sunk. If DuPont had not expanded early, other producers might have entered the market. After they entered, it might not have then made sense for DuPont to expand. The company's decision to expand did not affect the strategies available to the other players. They could build new plants if they wished, just as Entrant could still open a second cement plant. By making its investments early, however, DuPont changed the equilibrium strategies of the other players.

We can posit any number of steps that an incumbent might take—from signing a contract to buying equipment—for strategic advantage. The extensive form game we have set out here shows how many kinds of decisions can have strategic consequences. Even an act as simple as buying new equipment may represent a waste of resources, decrease consumer welfare, or both. Because merely changing the order of moves in an extensive form game can lead to a different course of play, any action that affects the sequence in which parties act can, in principle, have anticompetitive implications.

None of this, however, suggests that the scope of antitrust liability should be expanded or that the FTC was wrong when it refused to rule that DuPont's decision to increase its capacity was an "unfair method of competition." DuPont was, after all, the low-cost producer. The evidence suggested that it was still enjoying economies of scale. DuPont was successful in part because it could produce titanium dioxide more cheaply than anyone else. By committing itself to expand and inducing higher cost producers not to enter the market, DuPont's actions may have prevented others from spending resources inefficiently.

Even in competitive markets efficient firms may be the ones most likely to expand capacity. DuPont's decision may have had some foreseeable anticompetitive effects; indeed, internal memos predicting that the price of titanium dioxide would rise after these steps were taken are consistent with this conclusion. This possibility, however, was not sufficient to persuade the FTC to sanction DuPont and risk shielding less efficient producers from competition. This model suggests how actions, such as overexpansion, can be anticompetitive. Through the use of such models (and more elaborate variations on them), one can identify what actions can produce anticompetitive outcomes and how legal rules can change things for the better.

Subgame Perfection

We have been able to solve the preceding extensive form games by using backwards induction. From one perspective, our approach slights the Nash equilibrium solution concept. We could have used it to solve all the extensive form games that we have examined, because any solution found through backwards induction is also a Nash equilibrium. Moreover, many games that cannot be solved with backwards induction can be solved by using the Nash equilibrium concept. Using this concept, however, frequently proves difficult. In many extensive form games, there are multiple Nash equilibria. We have emphasized backwards induction in our discussion thus far because it is often the best way to isolate the one strategy combination that the parties are most likely to adopt.

Backwards induction, however, is not available in those cases in which the last player must move without knowing the other player's previous move. If we turn to the Nash equilibrium solution concept in such a case, we may need some means of identifying those Nash equilibria that are plausible and those that are not. We can get some idea of the problem by looking at a variation of the game involving Lender and Debtor that we examined earlier.

Assume that Lender must engage in extremely costly litigation in the event that Debtor defaults on the loan. Lender will prevail in the end, but the litigation will impose costs on Lender and Debtor of $125. Hence, instead of both enjoying a $5 profit from the transaction, each suffers a $120 loss in the event that Lender makes the loan, Debtor defaults, and Lender sues. (The assumption that litigation costs can exceed the stakes is not implausible, even with loans of substantial size. Litigation is like a contest because the chances of success turn in some measure on which party spends the most. Once one party spends money on litigation, the other party has an incentive to respond by spending a little bit more.) We illustrate this game in Figure 2.7.

We can solve the game in Figure 2.7 using backwards induction. Lender decides whether to lend money by asking what course can be taken after Debtor defaults. At this point, Lender does nothing. Lender prefers a payoff of −$100 to a payoff of −$120. Lender therefore infers that Debtor defaults when the loan has been made because Debtor knows that Lender will not sue after a default. Lender, believing that Debtor will default when a loan is made, refuses to make the loan in the first place.

We can also use the Nash equilibrium concept to solve this game. We do this by again looking at possible strategy combinations and asking whether each player is playing a best response given the strategy of the other. Recall that a strategy for a player spells out what that player does at each information set, even when it is an information set that will not be reached because of an action which that player has taken previously. In other words, even if Lender does not make the loan in

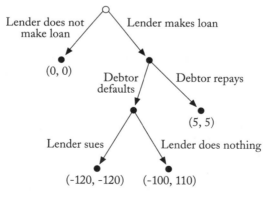

Figure 2.7 Debt contract with costly enforcement. *Payoffs:* Lender, Debtor.

the first place, Lender's strategy still reveals what Lender would do in the event that Debtor defaulted.

The logic of defining strategies in this way will become apparent as we proceed. It rests on the idea that we cannot analyze whether Lender's decision to lend the money makes sense unless we know what Lender would do when Debtor defaults. We implicitly used this notion when we solved the game in Figure 2.7 using backwards induction. (Lender adopts the strategy of refusing to make the loan and does not sue when Debtor defaults, whereas Debtor adopts the strategy in which Debtor defaults after the loan is made.)

This combination of strategies is also a Nash equilibrium. Lender's best response to Debtor's strategy of defaulting is not to lend the money in the first place. Lender is better off not making the loan and receiving a payoff of $0 than making the loan. When Lender makes the loan, Debtor defaults. When Debtor defaults, Lender does not sue. This leaves Lender with a loss of $120. Given that Lender's strategy in this proposed equilibrium is not to lend the money, Debtor's decision to default is a best response. Debtor cannot do any better by changing this strategy.

The Nash equilibrium concept, however, does not identify a unique solution to this game. It is also a Nash equilibrium for Lender to make the loan and to sue in the event of default, and for Debtor to pay the loan back. Lender is better off making the loan than not, given that Debtor repays. A payoff of $5 is better than a payoff of $0. Similarly, Debtor is better off repaying than defaulting, given that Lender sues. Debtor prefers a payoff of $5 to a payoff of −$120. A moment's reflection, however, will reveal that parties are unlikely to adopt these strategies. The solution rests upon Lender's threat to sue in the event that Debtor defaults and this threat is not credible. After Debtor defaults, it is no longer in Lender's self-interest to sue.

The Nash equilibrium solution concept can be refined to eliminate this second equilibrium. This equilibrium does not take into account Lender's incentives when Debtor actually defaults. The equilibrium rests on the idea that Lender sues when the game takes that course, but it does not seem plausible that Lender sues when Debtor defaults. As we have set the payoffs (which ignore, for example, reputational effects that might follow from a failure to sue), it is not in Lender's self-interest to sue.

A threat is ineffective unless the person who makes it will actually carry it out if called upon to do so. In a game such as the one in Figure 2.7, Lender will carry out a threat to sue Debtor to recover the loan only if, at the time Lender decides whether to sue, suing brings greater benefits to Lender than doing nothing. Any solution is suspect if it rests, directly or indirectly, on the assumption that Lender would bring a suit even though its costs would exceed the amount of the recovery.

We need a way to incorporate this idea into the Nash equilibrium concept. To put our ambition in game-theoretic language, we need a refinement of the Nash equilibrium solution concept so that we can exclude strategy combinations that, even though they are Nash, make implausible assumptions about the actions that the players would take, but do not in fact take under the proposed equilibrium. These are actions that are *off the equilibrium path*. This refinement of the Nash solution concept is known as *subgame perfection.*

A Nash equilibrium is subgame perfect if the players' strategies constitute a Nash equilibrium in every *subgame* of a game. A subgame is a move or a set of moves of an extensive form game that can be viewed in isolation. More formally, a subgame of a game in the extensive form is any part of a game that meets the following three conditions:

1. It begins at a decision node that is in an information set by itself.

2. It includes all the decision nodes and terminal nodes that follow it in the game and no others.

3. No nodes belong to an information set that includes nodes that do not also follow the decision node that begins the subgame.

The game in Figure 2.7 has two subgames. The first begins with Debtor's choice, the second with Lender's decision to sue or not to sue. The first subgame has two Nash equilibria. In the first, Debtor chooses to repay the loan and Lender sues when Debtor defaults; in the second, Debtor defaults and Lender does not sue. Focusing on this subgame does not allow us to eliminate any of the Nash equilibria in the whole game. But subgame perfection requires that the candidate solution be Nash in every subgame, so consider the subgame that begins when Lender chooses whether to sue. This subgame has only one Nash equilibrium: Lender does not sue.

To be "subgame perfect," a solution to the entire game set out in Figure 2.7 must be one in which Lender never adopts a strategy in which Lender would sue were Debtor to default. The proposed solution in which Lender makes the loan and Debtor repays it depends on Lender's suing if Debtor were to default. Hence, it cannot be a subgame perfect equilibrium. The only equilibrium that is subgame perfect is precisely the equilibrium that is reached through backwards induction.

As we have seen, the game in Figure 2.7 could be solved through a straightforward application of backwards induction. Backwards induction, however, is not available in games that have information sets with more than one node. Backwards induction depends on knowing the last move of the game. This may not be possible if a player does not know what moves have already been taken when the time comes to move. Subgame perfection is a useful refinement of the Nash equilibrium concept. Like backwards induction, it eliminates those solutions that rest upon threats that are not credible, as well as other implausible actions off the equilibrium path; unlike backwards induction, however, it is still available when information sets contain more than one node.

There are three complications that one might introduce into our model of the interaction between Lender and Debtor. The first two give formal content to the ideas of vengeance and retaliation. First, Lender may not make decisions entirely on the basis of monetary costs. To the extent that Lender suffers psychological harm or loss of esteem among peers from failing to recover a loan, we need to alter the payoff structure of the game accordingly. Although litigation has a dollar amount that exceeds the size of the loan, these other costs associated with failing to recover the loan may make litigation the course that brings Lender the greatest benefits. This model, like those in Chapter 1, posits hard out-of-pocket costs as the only component of the payoffs, but nothing in the structure of the model requires this.

The second complication requires us to change the structure of the model. Lender may lend money to many different borrowers. To model the behavior of Lender, we may not be able to focus on a one-time game. Instead, we might want to imagine this game as part of a larger game in which Lender engages in the same transaction repeatedly. To the extent that this is the case, we must ask whether Lender will pursue litigation because of the effect of a failure to sue on interactions with future creditors. We show how reputation can be modeled as a repeated game in Chapter 5.

The last complication arises from the way in which the strategy space in this model is limited. This model does not allow for the possibility of settlement. We explore in Chapter 8 how, once one introduces the possibility of settlement, Debtor might be willing to settle in the face of Lender's threat to sue, making the threat once again credible. The virtue of putting a formal structure on the behavior of Lender is that it makes us confront the question of whether a threat is credible and, if it is, why.

In the rest of this section, we show how subgame perfection illuminates a problem in contract law that often arises when goods are sold and then fail to work as expected. The buyer asserts that the seller failed to deliver the goods that the seller promised. The seller, by contrast, will argue that the goods were as promised, but that the buyer did not use them properly. Disputes arise because much depends on both the seller and the buyer exercising care in making or using the goods that may not be visible to the other unless there is expensive litigation.

We can illustrate the problem by examining several variations on *General Foods v. Valley Lea Dairy*, a commercial law dispute typical of many that arise every year. In 1978, General Foods bought 40,000 pounds of dry milk from a dairy cooperative. The dairy delivered the milk in 9 separate lots. Although it had no explicit contractual obligation to do so, General Foods tested each lot and found that 1 was contaminated with salmonella. This discovery led General Foods to retest the 8 other lots. When no further evidence of salmonella was found, it used them in its milk chocolate.

At this point, General Foods ran a third round of tests. Because the milk was once again in liquid form, the test was more accurate than the earlier ones. Before the results of this test were available, however, General Foods sold the chocolate to several candy makers. The third round of tests eventually revealed that the 8 lots were tainted. All the candy made from the chocolate had to be destroyed. General Foods reimbursed the buyers of the chocolate for the losses they suffered and in turn demanded that the dairy make it whole. General Foods, however, was unsuccessful in its efforts to recover civil damages from the dairy that had sold it the contaminated milk.[2] Much in this case turned on its peculiar facts. Our interest is in the general way in which this problem should be approached.

When food products such as milk are sold, there is always the risk that they are contaminated. The parties, however, want to confront the danger that salmonella presents in a sensible way. The problem is similar to the ones we examined in the last chapter. Each party should use due care in manufacturing and processing the milk. The contract they write should ensure that both parties have the incentive to test in a way that is cost effective, and, when they do not reach an explicit agreement, the gap-filling rules of contract law should try to give them the right incentive. If the laws do not do this job, we run a risk analogous to the one we saw in the example of the debt contract at the outset of this chapter. Rather than entering into a contract and taking advantage of the skills of another, a company such as General Foods may bring the operation inside the firm, where it will do the task less efficiently than the dairy could if it had the right set of incentives. If a legal regime can give the dairy the right incentives, everyone could be made better off.

Consider a variation on the facts of *General Foods*. A food processor that uses dry milk in various products it sells must decide whether to process its own dry milk or to buy it from a dairy. If it processes the milk itself, the firm can cheaply monitor whether the milk is contaminated. The workers who inspect its other food processing operations at the same site can inspect the initial drying of the milk with little added cost. A dairy can dry the milk more cheaply in its more specialized plant located some distance away. If the dairy dries the milk, however, the food processor must trust the dairy to hire people to monitor the initial drying process and run the first set of tests.

The dairy may perform two kinds of tests—either high or low. The high test is more expensive but also more accurate. The food processor runs a second set of tests after the dry milk is delivered. It has a choice of three kinds of tests: high, medium, or low. As before, the more expensive tests are more accurate. We can illustrate this problem as the extensive form game in Figure 2.8a.

Processor moves first and decides whether to buy the milk on the market. Dairy then decides on the kind of testing it will do, and then

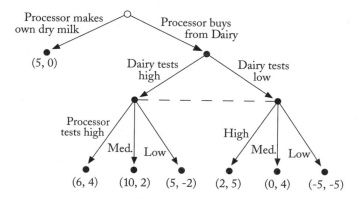

Figure 2.8a Sale without warranty. *Payoffs:* Processor, Dairy.

Processor decides on its tests. Even when there is no enforceable contract between it and Processor, Dairy incurs some added costs from using the cheaper test because of the loss to its reputation and potential lawsuits from third parties. Whether these costs are worth incurring turns on what kinds of tests Processor runs. As in the game in Figure 2.1, we draw a dashed line between the two possible decision nodes of Processor because they are both in the same information set. At the time when it must make its decision, Processor does not know whether Dairy has used the high or the low test and hence does not know whether it is at the decision node on the left or the right.

We cannot reason backwards move by move as we could in the games involving Lender and Debtor. If Processor knew that it was at the decision node that arises after Dairy decides to test high, it would test medium. If Processor finds itself at the right node, however, the outcome is different. If Dairy tests low, Processor is better off testing high. We cannot reason backwards from Processor's last move because it depends on what Dairy did in its previous move.

We can, however, solve this game by a different kind of backwards reasoning. We can isolate one part of the game—the subgame that begins after Processor decides to have Dairy process the milk—and solve it before solving the game as a whole. If we ignore Processor's initial move (the decision of whether to process its own milk or buy it from the dairy), we have a game that can stand independently. The single node at which Dairy must decide on the kind of test it wants to use could be an initial node of a free-standing game. We can solve the larger game by solving this subgame and, with that solution in hand, use backwards induction to solve the game as a whole.

The subgame that begins with Dairy's first move is one that we can easily capture in the normal form, because, after that point, each player must make its move without knowing what the other has done. We can illustrate the normal form of this game with a three-by-two bimatrix, shown in Figure 2.8b.

There are two solution concepts that we can use to find the likely course of play in this subgame. First, we can note that this subgame has only one Nash equilibrium, the strategy combination in which Dairy tests low and Processor tests high. If Dairy is going to test low, Processor is better off testing high and enjoying a return of $2, rather than testing medium or low and receiving a return of $0 or −$5. If Processor is going to test high, Dairy, of course, has no incentive to deviate either. It prefers receiving $5 from testing low to $4 from testing high. This subgame has no other Nash equilibria.

	Dairy's Level of Care	
	High	Low
Processor's Care — High	6, 4	2, 5
Processor's Care — Medium	10, 2	0, 4
Processor's Care — Low	5, -2	-5, -5

Figure 2.8b Sale without warranty (subgame after Processor decides to buy). *Payoffs:* Processor, Dairy.

We can also solve this game through the successive elimination of dominated strategies. Processor will never play low because the strategies of high and medium dominate it. Dairy will observe that, so long as Processor never moves low, it is better off moving low rather than high; hence, Dairy will move low. Processor, recognizing that Dairy will move low, will realize that it is better off moving high. Thus, the solution to the subgame is one in which Processor moves high and Dairy moves low.

Once we determine how the subgame is likely to be played if the players were to reach it, the solution to the game as a whole becomes clear. Processor knows that it will enjoy an expected payoff of only $2 if it decides to buy the dry milk from Dairy, but that it will enjoy a payoff of $5 if it makes the dry milk inside the firm. Because Processor will find it in its self-interest to play high whenever it decides to buy milk from Dairy, we must reject any proposed solution that rests upon Processor's adopting some different strategy if the game proceeds along that path. The game as a whole has several Nash equilibria, but of these only one seems plausible, and this strategy combination is the unique subgame perfect Nash equilibrium of the game. This equilibrium is the one in which Processor makes its own dry milk, Dairy tests low—when given the chance to move—and Processor tests high when it buys milk from Dairy.

Up to this point, we have been considering the problem between Processor and Dairy as it would exist if the milk were sold without a warranty. The outcome that has the greatest joint payoff for both players is that in which Processor buys the milk and tests medium, while Dairy tests high. Unless the parties take actions to perturb the payoffs, however, they will end up far from this outcome. One way to perturb

Dairy's Level of Care

		High	Low
Processor's Care	High	6, 4	5, 2
	Medium	10, 2	3, 1
	Low	5, -2	-2, -8

Figure 2.8c Sale with warranty (subgame; with contract). *Payoffs:* Processor, Dairy.

the payoffs is to provide a schedule of damages that Dairy owes Processor when the milk proves to be tainted. Contract damages do not change the joint payoffs, but they change the way they are distributed between the parties. Contract damages work in the same way as tort damages.

As we showed in the last chapter, there are many schedules of damages in a simultaneous move game that induce both parties to adopt the course that is most mutually beneficial. One such schedule gives us the subgame that begins after Processor decides to buy the milk, under which Dairy must pay damages of $3 any time it tests low.[3] The payoff transformations brought about by this damages schedule are shown in Figure 2.8c.

The unique Nash equilibrium of this subgame is one in which Processor moves medium and Dairy moves high. When Dairy moves high, Processor's best response is to play medium and enjoy a payoff of $10, rather than play either high or low and receive a payoff of $6 and $5 respectively. Dairy will test high if Processor tests medium, preferring a payoff of $2 to a payoff of $1. Hence, this strategy combination is Nash.

We can also check that no other combination is Nash. As noted, Processor's best response to Dairy's strategy of high is medium. Any other combination in which Dairy tests high cannot be a Nash equilibrium because Processor's strategy would not be a best response. If Dairy plays low, Processor's best response is high. The strategy combination of high on the part of Processor and low on the part of Dairy is not Nash because Dairy's best response to Processor's strategy of high is to play high as well.[4]

Given that in the subgame the strategy combination in which Dairy tests high and Processor tests medium is the only Nash equilibrium, we know that the only subgame perfect equilibrium to the game as a whole is the one in which Processor and Dairy adopt these strategies. If Dairy tests high when given the chance, Processor's best response in the game as a whole is to buy milk from Dairy rather than process the milk itself. Given the actions that Dairy will take in the subgame, Processor enjoys a payoff now of $10 if it buys milk from Dairy, rather than a payoff of $5 if it makes the milk itself.

This game has other Nash equilibria, but they rest on implausible assumptions about play off the equilibrium path. Consider the following combination of strategies: Dairy adopts a strategy of low in the event that Processor buys Dairy's milk. Processor in turn manufactures its own milk and tests high when it buys milk from Dairy. This equilibrium is Nash. Given that Dairy is going to test low, Processor's best response is to make its own dry milk. When Processor makes its own milk, it earns $5, but if it buys from Dairy and Dairy tests low, Processor can earn no more under this schedule of damages than $5. Dairy has no incentive to switch from a strategy of testing low. Because Processor is going to make its own milk, Dairy receives the same payoff of $0 regardless of how it tests when it sells the milk. Each player adopts a best response given the strategy of the other; hence, this combination of strategies is a Nash equilibrium.

This equilibrium, however, is not a plausible solution to the game. A combination of strategies in which Processor makes its own milk and Dairy tests low is Nash only because, in equilibrium, Dairy does not actually have a chance to move. Dairy would not in fact test low if Processor were to buy the milk from it. Under the new payoffs, testing low is no longer in Dairy's interest. Testing low, after all, is a dominated strategy and a player will not play a dominated strategy if given a chance to move. Moreover, the other player will act on that assumption.

Our solution to the game must take into account the incentives of the parties at every possible decision point. The technique of focusing on the play in each subgame to solve the game as a whole is useful because it eliminates any Nash equilibrium that rests upon players' taking implausible actions that are off the equilibrium path. Once we find that a particular combination of strategies is the unique subgame perfect Nash equilibrium, we have solved the game. For this reason, we can predict that, in this game, Processor should adopt a strategy of buying from Dairy and testing medium. Dairy should adopt a strategy of testing high in the event that it sells the milk. There are no other

Nash equilibria in which the players behave rationally both on and off the equilibrium path.

In this example, the parties wrote a contract that changed the payoffs in three cells of the payoff bimatrix. Most contracts do not have elaborate schedules of damages. Moreover, it is not possible for the background legal rule to provide such details for every possible commercial transaction. As we saw in the last chapter, however, it is possible to state in general terms obligations to pay damages that give everyone the right incentives. For this reason, the parties may be able to write a contract in which they adopt a general rule that transforms the payoff structure in a way that ensures that both parties act with the interests of the other in mind.

A crucial question—and one upon which we focus in the next two chapters—is how to formulate a general rule that can work even in a world in which information is hard to come by, both for the parties and for the courts called upon to enforce the rule. In the game involving Dairy and Processor, for example, we would want to explore whether we could induce optimal behavior with a schedule of damages that was tied to whether the milk was tainted, not to the kind of test that Dairy ran. It is much easier for a court to determine the former than the latter.

The problem that parties face when they enter into their initial contract is similar to the one facing a lawmaker who is devising tort rules and trying to minimize the social costs of accidents. If we make strong assumptions about the rationality of the players and the information available to both the players and the courts, many different rules are possible that transform the payoff structure from what it would be if the buyer had no cause of action against the seller and that also give each party the right set of incentives. A lawmaker, however, needs to design rules that handle those cases in which the parties do not explicitly take a particular contingency into account.

The Uniform Commercial Code governs a contract for the sale of goods and applies to cases such as *Valley Lea.* Two of its provisions (both of which the parties can waive if they choose) are relevant for us. The first provision is the implied warranty of merchantability. A merchant seller such as Dairy promises that its goods pass without objection in the trade under the contract description. If they do not, the seller must make the buyer whole. The second is the provision that requires the buyer to mitigate losses. A seller is liable only for those losses that "could not reasonably be prevented."[5] Under this rule,

Dairy is liable only for the losses Processor suffers if Processor exercises reasonable care and takes those steps that are cost effective.

As our discussion from the last chapter suggests, such a rule gives the players the incentives to exercise care, provided that we make a number of assumptions about such things as what losses could not be "reasonably" prevented. We should not, however, end the inquiry there. To say that this rule is efficient under this set of assumptions is very far from saying that it is the only rule that is efficient, given these assumptions. Recall that any extensive form game can be represented as a normal form game; and, as we saw in the last chapter, an infinite number of civil damages rules exist that give the players the right incentives. In the last chapter, we worried about whether a rule existed that made the fewest rationality assumptions. This concern, however, should not be our only one. Perhaps of even greater importance is whether a particular rule is informationally parsimonious. We may prefer a simple rule—one that parties can understand and that we can enforce—that gives us results that come close to the optimum, even if it would not work as well under ideal conditions.

The common law embraces the idea that parties are liable for expectation damages, subject to a duty to mitigate on the part of the innocent party. This rule itself is not perfect,[6] but even this rule, which requires a court to determine how much a party would have received if the contract had been performed, may be too informationally demanding. When parties draft their own damage provisions, they often opt for ones that are considerably more simple.[7] They often provide for fixed damages in the event of breach, and many items that would fall within the scope of expectation damages, such as consequential damages, are routinely excluded. The game with Processor and Dairy in this chapter showed how a simple damage schedule might work. In the next chapter, we turn to the considerably more difficult problem of showing how we can explicitly take account of informational problems in fashioning legal rules.

Summary

In this chapter, we have shown how we could capture as extensive form games situations in which individuals interact with each other over time. As in the last chapter, we have shown how we could model legal rules providing for civil damages as transformations of the payoffs. The payoff to one player under each strategy combination rises by the same amount that another player's falls. Once again, legal rules often ensured that the likely solution to the game was one that was in the joint interests of the players by changing payoffs that were off the equilibrium path. The dynamic character of these games, however, introduced a new problem.

In the tort problems that we examined in the last chapter, what mattered was that players internalize the cost of the decisions they make. When the other player makes decisions at the same time, little else matters. When players interact over time, however, what matters is not so much that a player internalizes all the costs of a decision as that a player always has an incentive to make decisions that keep the game on the path leading to the socially desirable outcome. The structure that the extensive form game imposes on problems in contract law and other problems involving long-term relationships between parties provides an opportunity to look at such doctrines as the parol evidence rule, conditions, and the Statute of Frauds, all of which have the effect of attaching consequences to actions that take place off the equilibrium path of the game.

The structure of the extensive form game and the idea of subgame perfection give us a way to distinguish credible threats from ones that are not. As we saw in the game involving Incumbent and Entrant, regulatory regimes such as antitrust must be sensitive to the frequently subtle ways in which parties can act so as to change the incentives they and others will have at subsequent points in time—and thus convert noncredible threats into credible ones. In addition, this structure provides a way of looking at regulatory regimes in general, and it focuses on the dynamic consequences of imposing a new regulatory regime.

We can illustrate this problem of dynamic consistency by considering a weakness in a one-time tax amnesty law that is coupled with the promise that another such law will never be passed. The ambition of the law is to force tax evaders to come forward, but also to convince people that tax evaders in the future will not be treated so generously. Such a law may not work because the promise never to institute another amnesty (essential to deterring people from future tax evasion) may not be credible. A regulatory regime, like the structure of a contract between private parties, must be dynamically consistent over time.

A policy in which one offers amnesty on only a single occasion can make sense only if one can explain why offering amnesty again at some later time would not appear at least as attractive. One must look not only at the play of the game as a whole, but at the play of the subgame that arises after the tax amnesty takes effect, people once again evade taxes, and the legislature decides whether to keep its promise not to provide another amnesty. The initial resolve of the legislature is credible only if it will be in the interest of the legislature to keep that resolve at a later point in time. If the legislature is not able to tie its hands, the promise never to have another amnesty will not be credible.

The same problem can arise in the context of litigation. In a Chapter 11 reorganization, for example, the debtor often requests an extension of the period in which the debtor has the exclusive right to propose a plan of reorganization. The court often grants the debtor's request for an extension, but asserts at the same time that it will not grant any additional extensions. Creditors are hurt by the passage of time because of the time value of money, but, except for the passage of time, nothing may change if no plan of reorganization is agreed upon before the period expires again. The court may find itself in exactly the same position when the debtor requests another extension of the exclusivity period. In such a case, the court's initial statement that it will not make another extension may not be believed. To model extensions of the exclusivity period, we must take the problem of dynamic consistency into account.

Bibliographic Notes

The extensive form game.   The concepts of the extensive form game and subgame perfection are the basic elements of modern game theory. Kreps (1990b) provides a good formal introduction to the extensive form game.

The extensive form game and the debt contract.   The idea that parties write contracts and devise mechanisms to ensure that each has the incentive to cooperate with the other is well known. Kronman (1985) is a good example. Kreps (1990b) as well as Fudenberg and Tirole (1991a) explores the potential weaknesses of backwards induction and subgame perfection as solution concepts.

Market preemption and strategic commitment.   The literature on the value of commitment starts with the classic book by Schelling (1960). Applications to entry and capacity decisions are contained in Dixit (1979) and Spence (1977). The titanium dioxide industry is discussed in Ghemawat (1984).

Backwards induction and subgame perfection.   Selten (1975) pioneered the concept of subgame perfection and other refinements of the Nash equilibrium solution concept. Kreps (1990a) also has a good introduction to subgame perfection. Fudenberg and Tirole (1991a) explores the basic principles at work in the extensive form game. Dynamic consistency is discussed in Kydland and Prescott (1977).

CHAPTER 3

Information Revelation, Disclosure Laws, and Renegotiation

In the first two chapters, we looked at situations in which everyone was completely informed, except possibly about what decisions the other player had made. We now examine games of incomplete information, situations in which one player possesses knowledge that the other does not. This informational asymmetry itself can affect the way each player behaves; and legal rules can play a large role in determining how parties share information with each other. Indeed, many important legal reforms have focused on information and whether and how it is conveyed. Laws, for example, may mandate disclosure of information. Those who acquire more than 5 percent of the stock of a publicly traded firm must disclose their interest to other investors.[1] A company that intends to close a plant may have to give advance notice of the closing to its employees.[2] In the case of real estate sales, the doctrine of *caveat emptor* is giving way to laws requiring sellers to disclose whether the basement leaks or the neighbors are noisy. In other cases, the government limits the transfer of information to prevent discrimination or protect rights of privacy. Laws exist, for example, that make it illegal for an employer to inquire whether an applicant is disabled. Prospective students at a federally funded educational institution may not be asked their marital status. In this chapter and the next, we explore the kinds of effects that rules governing information may have on the way people interact with one another.

The following is an example of a problem arising from asymmetric information. A buyer knows something about a piece of land that the seller does not. The seller is a farmer and the would-be buyer is a geologist. The knowledge is whether the land has valuable minerals on it. The person who lacks knowledge, the farmer, in our example, must act notwithstanding the uncertainty. That person, however, does not act arbitrarily. First, a person can draw on previous experience and weigh the different possibilities. The farmer might not know whether there is oil or other minerals on the land, but nevertheless might have some sense of the probabilities. A farmer in Texas might think that there is one chance in ten that there is oil on the land, but a farmer in Illinois might put the chance at only one in a hundred. Second, the person who lacks information can draw inferences from the way another person acts. A farmer who starts with the belief that the land is unlikely to have oil on it might think it more likely that the land contains oil if an oil company geologist comes and asks if it is for sale.

When one person has information that another does not, this asymmetry itself affects how both parties behave. Farmers will want to know the identity of their prospective buyers. Potential buyers will, to the extent possible, conceal information (such as their training in geology) that tends to increase the price. Equally important from our perspective, the kinds of legal rules that are in place affect the kinds of inferences that parties can draw. In this chapter, we are concerned with how two parties behave when the informed party has the ability to convey crucial information to the other if that party chooses to do so. We then go on to examine situations in which both parties are equally well informed, but neither has the ability to convey to a court what is known. We postpone until the next chapter the case in which again only one player is informed, but that party has no ability to convey the information to the other directly. Anything the other player learns is learned by drawing inferences from the actions that the first player takes. Before we can confront any of these problems, however, we must introduce a new solution concept.

Incorporating Beliefs into the Solution Concept

In this chapter, we use the extensive form game to model the interactions between parties when one has information that the other does not. To do this, we need to develop a new solution concept, known as the *perfect Bayesian equilibrium*. This solution concept builds on an idea that we already encountered in our discussion of iterated dominance in Chapter 1. When we examined games in which we used the repeated elimination of dominated strategies, we had to posit the beliefs that each player would have. We found a solution by positing that each player believes both that the other player will not play dominated strategies and that the other player shares these same beliefs. When we encounter situations in which parties are incompletely informed, we must also identify ideas that we can use to predict the beliefs (and thus the actions) of the players. These ideas are largely based on two principles. First, rational players should change their beliefs in light of the actions that other players take; and second, they should act in a way that is consistent with their beliefs. Moreover, in an equilibrium, the beliefs of the players should be consistent not only with their own actions, but also with those of other players.

In the previous chapter, we developed the concept of subgame perfection, the idea that the actions of the players were Nash, not only in the game as a whole, but also in every subgame. We are using the idea of "perfection" in a parallel sense here. One looks not at whether actions are optimal in subgames, but rather at whether actions are optimal given the beliefs of the players. A perfect Bayesian equilibrium is "perfect" in the sense that the actions are optimal given not only the actions of the players, but also the players' beliefs. A proposed solution is suspect if it requires one player to have beliefs that are inconsistent with the actions that player takes or the actions that other players take. A solution to a game should not assume that a party harbors such beliefs, just as a solution should not assume that a player consistently takes a course of action that is less than optimal given the actions of the other players.

We can illustrate the intuition behind these ideas by recalling the movie *The Maltese Falcon*. One of the principal figures in the movie is Kasper Gutman, played by Sydney Greenstreet. He spends seventeen years tracking down a gem-encrusted statue of a falcon that the knights of Malta had once offered in tribute to the king of Spain. He finally finds it in Istanbul in the hands of a Russian general named Kemidov. The statue has been covered with black enamel and appears to be a curiosity of only modest value. Greenstreet tries to buy the apparently worthless statue, but Kemidov refuses to sell it. Two of Greenstreet's confederates (played by Peter Lorre and Mary Astor) then try to steal it from the general. At the end of the movie, they discover that the statue they stole was only an imitation that the Russian general had substituted for the original after Greenstreet had offered to purchase it.

Greenstreet, of course, had not told the Russian general that the statue was gem-encrusted, but the general had inferred from Greenstreet's eagerness to buy the statue that it was of great value. Greenstreet made the mistake of bidding too much. As Peter Lorre says to Greenstreet in exasperation in one of the film's best moments, "You!

It's you who bungled it! You and your stupid attempt to buy it! Kemidov found out how valuable it was. No wonder we had such an easy time stealing it."

If we were to model the interaction between Greenstreet and Kemidov as a game, Greenstreet's strategy space might be the offers he could make. Kemidov's strategy space is whether to accept or reject any offer. A solution for this game, however, should take into account Kemidov's assessment of the chance that the falcon he possesses is the genuine, gem-encrusted Maltese Falcon and how he changes that assessment in light of the offer that Greenstreet makes. Kemidov will accept only an offer that exceeds the value to him of keeping the falcon, given his beliefs about the chances that the statue is gem-encrusted. Moreover, the beliefs that Kemidov has about the value of the falcon must be updated in light of the offer that Greenstreet makes. For example, if Greenstreet appeared and offered $100,000 for a statue that Kemidov had previously thought was worth only $10, Kemidov would reevaluate his original assessment of the worth of the statue. He would infer not only that it was worth more than $10, but maybe that it was worth even more than $100,000. After all, Greenstreet would make such an offer only if he knew something about the statue that Kemidov did not. Greenstreet has every incentive to offer Kemidov less than the statue's true value.

Our model of the game would have to begin with Kemidov's initial assessment of the likelihood that the statue was something other than an uninteresting ceramic figure of a bird. Kemidov starts with the belief that the chance that the statue is valuable is quite low. (We know this because he does not think it worthwhile to inspect the statue more closely or have it appraised before Greenstreet comes on the scene.) In addition to this background assumption, we want to specify how Kemidov should update his beliefs in response to Greenstreet's offer. There are essentially two kinds of offers that Greenstreet can make: offers that he would make for the statue if it were gem-encrusted, and offers that he would make regardless of whether the statue were gem-encrusted or a simple ceramic figure.

Let us assume for the moment that, if the statue were gem-encrusted, Kemidov would place a higher value on it than Greenstreet would. In other words, if Kemidov knew that the falcon were gem-encrusted, he would not sell it. What can we say about the likely outcome of this game? Given that Kemidov would not sell the falcon if it were gem-encrusted, Greenstreet would be able to buy it from him only if he made an offer that he would be willing to make (and Kemidov knew he was willing to make) even if the statue were exactly what it appeared to be. The likely course of play cannot be one in which Greenstreet makes an offer that he would make (and that Kemidov knows he would make) only if the statue were gem-encrusted.

In any equilibrium, Kemidov's beliefs have to be updated in light of Greenstreet's offer, and Kemidov must act optimally, given those updated beliefs. If the offer is one that Greenstreet would make only if the statue were gem-encrusted, and if Kemidov knew this, Kemidov should no longer believe that the statue is very likely a ceramic figure of little value. Greenstreet "bungled it," in Lorre's words, because he offered too much and thus unintentionally conveyed information about the value of the statue.

The equilibrium in a game in which a player has private information tends to be one of two general kinds. First, there is a solution in which the outcome is the same regardless of the information the other player has. (Greenstreet makes an offer that he is willing to make regardless of whether the statue is gem-encrusted, and Kemidov is willing to take such an offer.) Such a solution is a *pooling equilibrium.* The offer in such a case communicates no information—Kemidov assesses the offer in light of his own initial beliefs about the falcon and the chance that it is anything other than it appears.

The second possibility is one in which a player is better off taking an action, even though the action communicates information. If, contrary to what we have been assuming, Greenstreet puts a higher value on the falcon than Kemidov if it is gem-encrusted, but not otherwise, the outcome might be one in which Greenstreet makes an offer only if the falcon is gem-encrusted, and the offer would be accepted. Such a solution is a *separating equilibrium.*

The Perfect Bayesian Equilibrium Solution Concept

We want to formalize the basic ideas that we discussed in the last section. We do this by setting out the perfect Bayesian equilibrium concept in the context of an extensive form game. We use another variation on *Valley Lea,* the case involving the dairy and the food processor that we discussed in the last chapter. The word "Bayesian" in the name incorporates the idea that uninformed players put probabilities on different events and then update them using Bayes's rule when other players take actions that convey information. (In probability theory, Bayes's rule provides a means to capture formally the way rational people should update their beliefs in the wake of new information.)

The word "perfect" reflects the idea that beliefs and actions have to be consistent with each other.

The intuition behind this solution concept, as we suggested in our example drawn from *The Maltese Falcon,* is straightforward. A player who is uncertain about what another player has done nevertheless has beliefs, based, perhaps, on previous experience. In addition, these beliefs are updated in light of new information. A solution to a game should take these beliefs and a player's ability to update them into account. As stated, the beliefs and the strategies of the players should be consistent with each other.

A proposed solution to a game is suspect if it depends on one of the players having beliefs that are inconsistent with the actions that the players take in equilibrium, or if it requires players to take actions that are inconsistent with their beliefs. We can test whether a proposed equilibrium is a perfect Bayesian equilibrium in much the same way we tested for a Nash equilibrium. We ask whether, in the proposed equilibrium, a player's actions are a best response, given that player's beliefs and the actions and beliefs of the other players.

Assume that Dairy must decide whether to use dry milk itself or sell it to Processor. Whether it uses the milk itself or sells it, Dairy is exposed to potential tort liability if the milk is tainted. The benefits of selling the dry milk to Processor therefore turn both on how much testing Dairy does and how much testing Processor does. Dairy moves first and has three choices. It can decide to process the milk itself and enjoy a profit of $3. If Dairy decides to process the milk itself, Processor does not get to move, and it receives a payoff of $0. If Dairy sells the milk, it must either test low or test high. Processor in turn must test low or high. If both test high, the tests are redundant. They will both spend so much on testing that each will enjoy profits of only $1. If Dairy tests high and Processor tests low, Dairy still enjoys profits of only $1, but Processor saves on the costs of testing and earns profits of $3. If Dairy tests low and Processor tests high, their positions are reversed—Dairy enjoys profits of $5, but Processor enjoys profits of only $1. If both test low, the ensuing tort liability reduces both their profits to $0. Figure 3.1 illustrates the extensive form of the game when Processor knows how much testing Dairy has already done.

In this game, Processor knows what kind of test Dairy has run when it moves. This game has multiple Nash equilibria. One arises from the strategy combination in which Dairy sells and tests low, and in which Processor tests high in the event that Dairy tests low, and tests low in the event that Dairy tests high. This solution is Nash because both

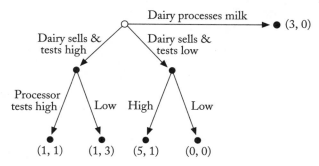

Figure 3.1 Sale v. production in the firm (Processor knows whether Dairy tested high or low). *Payoffs:* Dairy, Processor.

Dairy and Processor are playing a best response, given the strategy of the other. It is, of course, easy to see that Dairy can do no better. Under this strategy combination, Dairy enjoys a payoff of $5, its highest possible payoff. Seeing that Processor will not want to deviate is similarly straightforward. When Dairy tests low, Processor cannot improve its position by deviating from the strategy of testing high. It is better off testing high (and enjoying $1) than testing low (and receiving $0).

Another strategy combination that is Nash is the one in which Dairy processes the milk and in which Processor tests low when Dairy sells the milk. Given that Processor moves low, Dairy is better off processing the milk itself. Because Dairy is going to process the milk itself, Processor cannot improve its lot by deviating from testing low—and instead testing high—in the event that it receives the milk.

As we saw in the last chapter, we can use backwards induction to solve this game instead of the Nash equilibrium concept. We focus on the player who moves last—in this case, Processor. If given a chance to move, Processor will find it in its self-interest to test low if Dairy tests high (and thereby receive $3 instead of $1), but to test high if Dairy tests low (and thereby receive $1 instead of $0). Hence, when Dairy decides on its move, it can predict how Processor will respond to its strategies of testing high and testing low respectively. With this prediction in hand, Dairy will choose to sell and test low (and receive a payoff of $5), rather than sell and test high (and receive a payoff of $1) or process itself (and receive a payoff of $3).

We are able to use backwards induction in this game because Processor knows what kind of test Dairy ran when it moves. (The extensive form in Figure 3.1 tells us this because the left- and right-hand nodes at Processor's moves are not connected with a dashed line, indicating that they are in separate information sets.) Solving the game, however, is less straightforward if Processor does not have this knowledge. Consider the same game again, except that Processor does not know what kind of test Dairy ran. This new game is set out in Figure 3.2. The dashed line connecting the two nodes at which Processor might find itself on its move shows that Processor does not know whether it is at the left-hand node or the right-hand node when it moves.

We cannot use backwards induction to solve this game. The players make their testing decisions independently of each other, if they test at all. In addition, this game has multiple Nash equilibria. Hence, we cannot rely simply on the Nash solution concept to solve this game either. We cannot use the idea of subgame perfection. There is no proper subgame other than the game as a whole. There is no information set that contains a single node, other than the initial node. One cannot isolate part of the game and posit the strategy choices of the parties once the game reaches that node.

Because knowledge of Processor's position is not built into the game, we have to focus on the beliefs that it has about its position. When we incorporate beliefs into the solution concept, we can eliminate some Nash equilibria as solutions to the game. We can make some plausible assumptions about these beliefs, about how actions of the players affect these beliefs, and about how they in turn affect the actions of the players. If, for example, Processor believes that there is a 50–50 chance that Dairy tests high, it should test low. Testing low gives Processor an expected payoff of $1.5, whereas testing high gives it a payoff of only $1.[3]

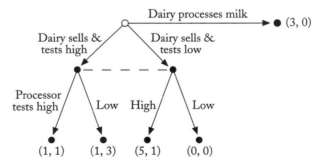

Figure 3.2 Sale v. production in the firm (Processor does not know Dairy's test). *Payoffs:* Dairy, Processor.

Any equilibrium in which Processor believes that Dairy is as likely to test high as to test low is one in which Processor must test low. Any other course is not rational for Processor given its beliefs. The Nash equilibrium concept requires that a player's strategy be optimal given the strategies of the other players. Once we incorporate beliefs into our solution concept, we also require that a player's strategy be consistent with that player's assessment of the probability of being at each node in the information set. These two ideas together are known as the requirement of *sequential rationality*.

As with the case of *The Maltese Falcon*, we must ensure that a player's assessment of the chances of being at any given node is consistent with the equilibrium action of the other players. Just as we do not think that players will embrace an equilibrium in which one player adopts a strategy that is suboptimal given the strategy of another, we should think that, in equilibrium, a player's beliefs are going to be consistent with the actions of the other player. Any equilibrium, for example, in which Processor believes that Dairy is as likely to test low as to test high, must be an equilibrium in which Dairy is in fact equally likely to test low as to test high in the event that it sells the milk. In equilibrium, the strategy choices of each player should be the best responses to the strategy choices of the others. The beliefs of the players should also be consistent with the strategy choices the other players have made.

In the game in Figure 3.2, there are multiple perfect Bayesian equilibria. Consider the case in which Dairy processes the milk itself, Processor tests low, and Processor believes that, if Dairy sells the milk, there is a 40 percent chance that Dairy tests high. This combination of strategies and beliefs forms an equilibrium. Dairy is behaving optimally given Processor's strategy of testing low. Processor's strategy is sequentially rational, given its belief that there is a 40 percent chance that Dairy will test high.[4]

There is nothing magical about Processor's belief that Dairy has a 40 percent chance of testing high. We would reach the same conclusion if Processor believed that there was a 35 percent or a 45 percent chance that Dairy would test high. Indeed, a perfect Bayesian equilibrium exists whenever Dairy processes the milk, Processor tests low, and Processor believes that there is more than a 33 percent chance that Dairy will test high. The belief that the chance of testing high is 40 percent, like the belief that it is 35 percent or 45 percent, is not inconsistent with Dairy's actions because Dairy never sells the milk in the proposed equilibrium. Processor never has any action that it can use to update its starting assessment, and this solution concept puts no restrictions on that starting assessment. This combination of actions, however, does depend on Processor's believing that there is at least a 33 percent chance that Dairy will test high. If Processor believed that the chance that Dairy would test high was less than 33 percent, Processor's best response, given this belief, would be to test high rather than low.

Another perfect Bayesian equilibrium in this game is one in which Dairy tests low, Processor tests high, and Processor believes that, in any case in which Dairy sells the milk, it will test low. Dairy's strategy is optimal because it is receiving the highest possible payoff. Processor is acting in a way that is sequentially rational given its belief that Dairy has tested low and given that Dairy has, in fact, tested low. Finally, Processor's belief is consistent with Dairy's equilibrium strategy of testing low.

Bayes's rule requires a player to update beliefs only in the wake of actions that take place in equilibrium. If an action is not taken in equilibrium, there is no new information that a player can use to update beliefs. Bayes's rule therefore does not place any constraints on a player's beliefs about actions that are taken off the equilibrium path. In the last chapter, we refined the Nash equilibrium concept in order to eliminate strategy combinations that rest on threats that are not credible. We can refine the perfect Bayesian equilibrium concept in the same way to ensure that a player's beliefs are plausible, even when those beliefs concern actions that never occur in equilibrium.

Consider again the equilibrium in which Dairy processes the milk itself and Processor believes that, were Dairy to sell the milk, there is again a 40 percent chance that Dairy will test high. Bayes's rule does not constrain Processor's belief because the case in which Dairy sells the milk is off the equilibrium path. Once an action is off the equilibrium path, nothing in the perfect Bayesian equilibrium solution concept constrains Processor's beliefs. Nevertheless, it seems implausible that Processor believes that Dairy tests high when it sells milk. Dairy always does better by processing the milk itself than it does by selling the milk and testing high. Dairy's strategy of processing its own milk dominates the strategy of selling and testing high. Processor is not likely to have a belief that requires Dairy to play a dominated strategy.

Processor's beliefs are likely to be molded by both Bayes's rule and the principle that people do not play dominated strategies. Processor should therefore infer that whenever Dairy sells the milk, it will test low. Once we posit that Processor has this belief, we can solve this game. The only perfect Bayesian equilibrium that survives this refinement is the one in which Dairy sells the milk and tests low, Processor tests high, and Processor believes that Dairy tests low when it sells the milk.

Verifiable Information, Voluntary Disclosure, and the Unraveling Result

The perfect Bayesian equilibrium with refinements is a tool that we can use to analyze how parties might draw inferences from the actions of other players. In this section, we want to examine these issues in the context of laws governing the disclosure of information, such as a law that requires a seller to disclose all known defects in a product. At the outset, we must recognize that there are different kinds of information. Some information is *verifiable;* that is, it can be readily checked once it is revealed. For example, the combination to a safe is verifiable information. The combination either opens the safe or it does not. Other information is *nonverifiable.* An employer wants to know whether a guard who was hired was vigilant. The employer might be able to draw inferences from some events, such as whether a thief was successful or was caught, but such information may not be available and, even if it is available, may not be reliable. Even a lazy guard may catch a thief, and a thief may outwit even the most vigilant guard.

Legal rules governing information have to take into account whether verifiable or nonverifiable information is at issue. When we analyze laws that require disclosure of information, we must confine ourselves to information that can be verified after it is disclosed. Moreover, we must limit ourselves to cases in which a court can determine whether a player possesses the relevant information. Courts cannot sanction parties for breaching a duty to disclose information if they have no way of telling whether a disclosure is truthful. Similarly, a court cannot sanction a party for failing to reveal information if it has no way of determining whether a party possesses the information in the first place.

At the outset, we want to focus on information that, once revealed, can be verified, both by the other players in the game and by a court that must subsequently apply the legal rule. We then examine situations in which the issue is slightly more complicated. We explore first an important principle in the economics of information known as the *unraveling result.* Consider a seller with a box of apples that has been sealed. The box can hold as many as 100 apples. The seller knows the number of apples in the box. The buyer does not know the number of apples in the box and has no way of counting them before the sale. The buyer, however, does know that the seller knows the number of apples in the box. After the buyer purchases the box and takes delivery, the buyer can count the apples. If the seller lied about the number of apples in the box, the buyer can sue the seller at no cost and recover damages. The buyer has no action against the seller who remains silent. Once we make these assumptions, we can conclude that all sellers will accurately and voluntarily disclose the number of apples in the box, no matter how few there are.

Focus on a seller with 100 apples. This seller will disclose. Because the information is verifiable and because of the legal remedy, the seller's disclosure will be believed and the seller will be able to charge the price for 100 apples. If the seller does not reveal the number of apples in the box, there is no way that the seller can persuade the buyer to pay the price of 100 apples. As long as the buyer believes that there is some chance that there are fewer than 100 apples in the box, the buyer will be unwilling to pay for 100. Hence, the seller with 100 apples discloses.

Now consider whether an equilibrium exists in which the seller has 99 apples and says nothing about how many apples are in the box. The seller would find remaining silent a good idea only if the buyer believed that there was some chance that there were 100 apples. The buyer, however, will not hold this belief in equilibrium because, as we have just seen, in equilibrium the seller with 100 apples discloses. If the seller remains silent, the buyer will infer that the seller has 99 apples or fewer. By remaining silent, however, the seller with 99 apples becomes lumped with sellers with very few apples.

This seller, knowing that the buyer will never pay the price of 100 apples, prefers to reveal that there are 99 and receive that price rather than one that reflects the chance that there may be fewer than 99 apples in the box. Because again the information is verifiable and the buyer has an effective legal remedy, the buyer will know that the seller is telling the truth. The seller with 98 apples goes through the same reasoning process, as does the seller with 97 apples, and so forth. This is the unraveling result. Silence cannot be sustained because high-value sellers will distinguish themselves from low-value sellers through voluntary disclosure. In the end, all sellers disclose their private information.

Whenever we examine any laws governing disclosure of information, we need to be aware of the possibility of unraveling. Cases involving the Fifth Amendment's privilege against self-incrimination illustrate this point. The Fifth Amendment to the United States Constitution provides that "[n]o person . . . shall be compelled in any criminal case to be a witness against himself." The privilege against self-incrimination is one of the most basic rights of criminal defendants. This right would be a limited one, however, if the failure to testify were to lead to an inference against the defendant. It was not until 1965, in *Griffin v. State of California*,[5] however, that the Supreme Court held that the self-incrimination privilege barred comment on the failure to testify. It was not until 1981, in *Carter v. Kentucky*,[6] that the Court held that the right required an instruction to the jury, on the defendant's request, that no inference be drawn from the failure to testify:

> A trial judge has a powerful tool at his disposal to protect the constitutional privilege—the jury instruction—and he has an affirmative constitutional obligation to use that tool when a defendant seeks its employment. No judge can prevent jurors from speculating about why a defendant stands mute in the face of a criminal accusation, but a judge can, and must, if requested to do so, use the unique power of the jury instruction to reduce that speculation to a minimum.[7]

Given the logic of unraveling—that someone with information will disclose it, rather than be subject to the inference that arises from the failure to disclose it when one can do so—the privilege against self-incrimination becomes meaningless unless steps are taken to prevent the adverse inference from being drawn. As the Court recognizes, however, there may be insurmountable limits on the ability of the legal system to prevent inferences from being drawn from silence. A jury instruction may make the problem worse because it may alert the jurors to their ability to draw inferences from a defendant's failure to testify. To be sure, we can infer that this jury instruction aids those who do not testify if defense lawyers regularly ask for such an instruction. Nevertheless, we may doubt, given the ability of the jury to draw inferences notwithstanding the instruction, whether the jury instruction is in fact "a powerful tool."

Because the person who holds favorable verifiable information has an incentive to reveal it, the allocation of the right or duty to inquire or disclose should not affect whether verifiable information is revealed. The idea that verifiable information may be disclosed voluntarily has important implications. It calls into question two standard legal approaches to revelation of information—inquiry limits and disclosure duties. We examine each in turn.

Inquiry limits attempt to prevent decisionmakers from obtaining information thought to be an inappropriate basis for making a decision. For example, the Americans with Disabilities Act bars an employer from asking whether an applicant has a disability and also bars preoffer medical tests.[8] A similar restriction applies regarding inquiries of applicants to educational institutions.[9] There are also limits on the kinds of inquiries that can be made of an applicant to rent or purchase a dwelling.[10]

Prohibiting questions about disabilities is a prominent form of inquiry limits, but others are common as well. Regulations implementing Title IX of the Education Amendments of 1972 forbid inquiry into the marital status of an applicant for employment at or admission to a school receiving federal funding.[11] Many states bar preemployment inquiries into religious or political affiliations for prospective public school teachers or for all prospective state employees.[12] Illinois bars inquiries into whether a prospective employee has filed a worker's compensation claim.[13] Inquiry limits are especially common under rules of evidence. The general protection for privileged matters—usually matters between attorney and client, physician and patient, or spouses, for example—is a form of inquiry limit, as are rules forbidding questions regarding a victim's prior sexual history in rape trials.[14]

Inquiry limits, however, may be ineffective unless there is some mechanism that prevents voluntary disclosure of the information. Barring an employer from asking whether an applicant has a disability might not affect whether the employer learns of the disability. When applicants who are not disabled also know the legal rule, they can disclose this information when it is verifiable.

As soon as the healthiest applicants disclose the results of their medical tests, the slightly less healthy ones may follow suit. The inability of the employer to require a medical test before making an offer may be irrelevant if applicants know that their employer wants the information, but cannot ask for it. They can simply volunteer it. Information problems can turn our usual intuitions upside down. We generally think that parties must know the legal rule for it to be effective. In this context, a legal rule barring inquiry might work best if the applicants did not know the legal rule. Knowledge of the legal rule might itself suggest the importance of the information to the employer.

The unraveling problem may give us some guidance about how inquiry limits might be made effective. Assume, for example, that the information is such that, if a job applicant volunteered it, the employer could not readily ascertain whether the applicant was telling the truth, but might be able to make that determination after the fact. (Applicants, for example, might volunteer that they never drank alcohol and had never had a drinking problem. Although the employer might have no ability to verify this information at the time the applicants made these statements, the employer might well learn that such statements were false at some subsequent time.) In such a case, a law that forbids inquiring might work effectively only if the law either provided some means to prevent applicants from volunteering the information or provided some means of keeping the employer from responding adversely upon discovering that an applicant had lied.

A rule that prevents the employer from retaliating against applicants who lied is another example of how a legal rule can change the way individuals behave even though it attaches consequences to actions that do not appear in equilibrium. Because applicants' assertions that they never drank would no longer be credible, applicants would no longer make them; or, if made, such assertions would no longer convey information. Once the employer cannot retaliate against the liar, the lies will not be made in the first place. A law forbidding employers from retaliating against workers who lied might be necessary to ensure that a law limiting an employer's ability to inquire into information operated effectively. It would also ensure that, after applicants were hired, they could disclose information, thus allowing the employer to accommodate their disabilities.

These problems with inquiry limits appear elsewhere as well. A law forbidding an inquiry into political affiliation seems to serve little purpose, as long as applicants for city jobs know that those hiring want employees from a particular political party. Applicants from that party will make their sympathies known, and those doing the hiring can draw an inference from silence. Rules limiting the transfer of verifiable information should be two-sided. In other contexts, two-sided bars are used. For example, it is unethical both for an attorney to condition an offer of settlement of a suit against a client on opposing counsel's agreement not to sue that client again[15] and for an attorney to accept such an offer.[16] As a general matter, however, statutes with inquiry limits are one-sided, and, in cases of verifiable information, such rules are potentially defective.

Laws requiring disclosure of information raise similar problems. An important issue in labor law concerns the employer's duty to convey information to the union. The Supreme Court held in *NLRB v. Truitt Manufacturing Co.* that an employer who claims financial hardship as a reason for not increasing pay must be willing to back up its claim.[17] Failure to provide the relevant information may support a finding that the employer has not bargained in good faith. As the Court noted, if a claim of financial hardship "is important enough to present in the give and take of bargaining, it is important enough to require some proof of its accuracy."

Strikes may arise as a result of private information. A union may be willing to incur the costs of a strike as a way of distinguishing between the high-profit firms that can afford a higher wage and the low-profit firms that cannot. If the high-profit firm loses more when there is a strike, it may be willing to settle earlier at a higher wage. The union could then infer that those firms that refuse to settle are not losing much from the strike because they are low profit.[18] If the legal rule had the effect of revealing the types of all firms at the outset, there would be no need for a strike.

Whether we can justify the rule of *Truitt* on this basis, however, is not obvious. As long as the private information was verifiable, low-profit firms would voluntarily disclose the information. High-profit firms that claimed low profits, but did not support the claim, would not be believed. The legal rule of *Truitt* would not affect the amount of information conveyed. With or without the rule, low-profit firms would disclose information and high-profit firms would not. *Truitt* would affect what high-profit firms said in the bargaining. They would have to be careful not to claim financial hardship, but this in itself would not make any difference. Unsupported statements of the high-profit firms would not be believed regardless of the legal rule. Justifying *Truitt* seems to require in the first instance confronting the question of why the unraveling result may not hold. (We might, for example, ask whether it is costly for low-profit firms to disclose financial information.)

There are other ways in which the unraveling principle informs our understanding of disclosure laws. For example, the Securities and Exchange Commission (SEC) has recently retreated somewhat from the strict disclosure requirements for issues that are not large and not widely held, in part because of the way the unraveling principle works in the absence of a legal rule. Sophisticated investors will infer bad news from an issuer's failure to disclose relevant financial information. Therefore, the incentives of the issuer would appropriately trade off the costs of information acquisition with the value of information.[19]

The unraveling result may also help to justify some rules that forbid the disclosure of information. Banks and bank examiners are banned from discussing bank examination reports publicly. Bank examiners gather information about a bank's condition and then use this information in their negotiations between a bank and regulatory authorities to fix problems before the risk of failure becomes severe. If banks were free to make reports public, banks would, on average, be made worse off because of unraveling. Without a law forbidding disclosure, the bank that receives a good examination report will want to make it public; it will raise the bank's stock price and reduce the probability of a run. Unraveling will begin, however, and depositors will in turn infer that any bank that does not reveal its report received a bad one. Although a bank in a relatively weak financial condition might have been able to work through its problems, it will not survive when depositors infer that there is a small risk of failure and then rush to take their money out. Disclosure, in other words, gives only a small benefit to good banks, does nothing one way or the other for bad banks (which will fail anyway), but harms banks that fall in the middle. Because of the unraveling principle, the law works only if limits are placed on a bank's ability to talk about a report, regardless of whether it is favorable.

Disclosure Laws and the Limits of Unraveling

The unraveling result depends on the ability of one player to infer the other player's information from that player's silence. Players are able to reveal only the information that they acquire. This linkage between acquisition and revelation of information may itself prevent unraveling. Unraveling may not occur (or will not be complete) if there is a chance that a player has never acquired the relevant information. In such a case, one will not be able to tell whether players are silent because they do not have the relevant information or because they have the information but do not wish to reveal it. The inability to draw inferences from silence in these cases might seem to justify some legal rules. Mandatory disclosure laws, however, may not work in this context. A law that mandates disclosure requires a court to distinguish players who withhold information from those who do not have it. If we need such a law only when the uninformed player cannot draw such a distinction, we must explain why the court is able to do something that a player cannot. The question, in other words, is whether a court can subsequently determine whether a player who claimed ignorance was telling the truth and inflict a penalty large enough to ensure that no one would have an incentive to lie.

Instead of a case in which both players know that one possesses the combination to a safe, we have a case in which one player may or may not have the combination to a safe, and the other player does not know either the combination to the safe or whether the other player knows it. A law requiring a player who knows the combination to a safe to disclose it cannot be enforced if a court lacks the means to determine whether a player knows the combination. Disclosure laws work only if a court can gain information that the uninformed player cannot.

Opposing litigants have the power to subpoena witnesses, compel testimony under oath—subject to criminal penalties for perjury—and discover documents. For these reasons, courts may be able to determine whether a party who fails to disclose information broke the law or simply did not have the information in the first place. Moreover, the sanctions that the court can impose may be sufficiently large to make disclosure laws effective. Nevertheless, this limitation on disclosure laws must be acknowledged.

We can examine the way this complication affects the unraveling result by returning to the example with the 100 apples. In this variation, the seller may or may not know how many apples are in the box. The buyer knows neither the number of apples in the box nor whether the seller knows how many apples are in the box. A court is able to determine after the fact, however, whether the seller had such knowledge at the time of the sale. In this case, the buyer is no longer able to draw the same inferences from silence. Silence may reflect the seller's ignorance rather than an unwillingness to reveal bad information. The ignorant seller, of course, is better off if a buyer can distinguish between those sellers who are ignorant and those who know how many apples are in the box. The ignorant seller, however, will not be able to prove the absence of knowledge to a buyer.

Unraveling also depends on whether the uninformed party knows what kind of information the other party has and might wish to conceal. In the absence of at least this much knowledge, a party cannot draw any inferences from silence. Some prospective home buyers may not realize that lead paint poses a danger. They will not infer that the house has lead paint from the silence of an owner of a house with lead-based paint. That sellers of houses free of lead paint would disclose the information does not help them. One may justify the SEC's decision to relax strict disclosure requirements only for issues that are not widely held, in part because less-sophisticated investors may not draw the correct inference from silence.

Close examination of the unraveling result may reveal other assumptions on which it depends. For example, unraveling took place in the example with the apples because everyone understood that, holding everything else equal, buyers prefer more apples to fewer apples. Unraveling might not take place if the preferences of buyers were not uniform. Assume, for example, that a box contains 100 balls. The balls are all either red, green, or blue. The seller knows their color and the buyer does not. Each buyer prefers some colors over others, but their tastes are not uniform. In such a case, the buyer may not be able to infer the color of the balls from the seller's silence.

Some kinds of information that are subject to legal regulation may not be subject to the same neat ranking as the number of apples in a box. Prospective job applicants who announce that they are teetotalers do not necessarily raise their stock in the eyes of a prospective employer, even if the information is true and can be verified. The employer does not want to hire anyone with alcohol problems, but the employer might not want to hire teetotalers either. The employer might believe that they are not sufficiently sociable. Note that the source or the accuracy of the employer's beliefs is irrelevant; what matters is that the employer holds them.

In the rest of this section, we examine mandatory disclosure laws as they apply to cases in which there is verifiable information, but in which, for the reasons we have just mentioned or for others, full unraveling does not occur. Disclosure laws typically require parties to reveal all the relevant information they possess. Sellers who sell pictures they know to be forged can be sued for fraud. Other rules require parties both to acquire and to reveal specific information. For example, a California law requires sellers of homes to fill out an extensive disclosure form covering appliances included and the extent to which they are operating, requiring disclosure of "significant defects / malfunctions" in walls, ceilings, and plumbing, as well as disclosure of environmental problems, flooding problems, and "neighborhood noise problems or other nuisances."[20] Other states have adopted similar legislation.

Some laws require both gathering specific information and disclosing information already known. The federal law governing corporate disclosure, for example, requires specific information in annual reports, 10-Ks, and proxy statements. In addition, federal law also requires firms that release any information about a transaction to reveal all the information about the transaction that a reasonable investor would find useful in determining whether to buy or sell the firm's security at a particular price.

As we have noted, such a legal rule works only if courts can determine whether a party knew the relevant information. Moreover, the existence of such a law itself affects whether someone becomes informed in the first instance. A seller may be less likely to test for lead paint if the results of any test must be disclosed to potential buyers. A law that requires a party to gather certain information and then disclose it circumvents this problem. Such a rule may also make it easier to understand the information and make comparisons between different players. A firm's annual report is easier to interpret if earnings must be calculated according to a standard accounting convention. Moreover, it is easier to compare one firm with another if the same information must be disclosed for both in the same format. Such laws, however, also bring with them a substantial cost. They may force a party to invest resources in gathering information that neither party cares much about.

One of the goals of laws governing the transmittal of verifiable information should be to induce efficient acquisition of information. It is to this question that we now turn.[21] Information sometimes brings benefits to one party to a transaction, but not to both. For example, in *Laidlaw v. Organ*, one of the parties spent a lot of money to become the first to know that the War of 1812 was over and that the British blockade of New Orleans was soon to be lifted.[22] This information, however, only gave the person who gathered it the ability to profit at someone else's expense. There is no (or at least very little) social benefit in knowing a few hours earlier than anyone else that a peace treaty had been signed many weeks before. Ideally, disclosure laws should discourage the acquisition of information in such cases. Disclosure laws that require parties to disclose all relevant information may have this effect.

When there is no obligation to disclose information and unraveling does not take place, information about whether the war is over has value to the person who knows it even though it has little or no social value. Because of its value to the individual, that individual will spend resources acquiring it. By contrast, if disclosure were mandatory, a party that spent resources gathering information would have to disclose it even if it were bad. The gain when the information is good is offset by the loss when the information is bad.

On balance, the informed person does no better than the uninformed one. Indeed, because the information is costly to acquire, the informed person does worse, and hence has no reason to gather the information in the first place. Because the information, by assumption, has no social value, a law requiring disclosure has a desirable effect. When information has no social value, a law requiring disclosure works better than one that imposes no disclosure requirement. Both, however, are preferable to a law that requires parties to gather and disclose such information. Requiring parties to gather information in such a case forces them to waste resources.

A law requiring disclosure only in cases in which information has no social value, however, may be hard to write. In addition, we always have to ask whether such a law can be implemented effectively in any case in which the unraveling principle does not apply. A duty to disclose makes sense only if a court is able to determine after the fact what a party, such as the buyer in *Laidlaw*, knew at the time of the transaction. Once we assume that a court can do this, however, it is hard to see why unraveling would not occur even without a disclosure law. Even if the buyer did not volunteer the information, it would be a simple enough matter for the seller to ask for, and insist on, a contract that would hold the buyer liable for any misstatements. Indeed, *Laidlaw* itself may illustrate the point. The court held that the buyer did not have to volunteer his private information; the court did, however, require a new trial on the question of whether the seller asked the buyer about war news and whether the buyer's answer was nonfraudulent.

We want to ensure that parties gather information when the social benefits of the information exceed the cost of gathering it, but doing this is hard when information brings a private benefit to one of the parties as well as a social benefit. We need to focus on both the social value of the information and the size of the private benefit that the party who gathers the information enjoys. A legal rule works effectively when it ensures that these are equal.

A straightforward model illustrates what is at work. A seller wants to sell a house. The seller can gather information about the house at some cost. The legal rule determines both whether the seller must gather the information and, if the information is gathered, whether it must be disclosed. We start with the simplest case—that in which the information affects the price at which the house is sold, but has no social value.

Let us assume that the value of a house is determined by whether its furnace is likely to need replacing in the next year. A house that needs a new furnace is worth only $180, whereas one that does not is worth $200. There is nothing to be gained, however, from knowing in advance that a furnace will need to be replaced. Whether a house needs a new furnace depends upon when the existing furnace was installed and how heavily it has been used. The buyer has no way of gathering this information. The seller, however, can acquire this information and convey it to the buyer by tracking down the original sales receipt for the furnace and the heating bills incurred since then. If the seller does not track the information down, the seller is in the same position as the buyer and knows only that the house is equally likely to need a new furnace as not. Half the sellers can track down the relevant information at a low cost and half at a high cost.

We can consider three different legal regimes. First, we consider a regime of voluntary disclosure, in which sellers are free to decide whether to find out if the furnace should be replaced and, if they do, are free to show their sales receipts and other relevant material to the buyer. Second, we consider a regime in which the seller is required both to gather the receipts and to show them to the buyer. Last, we consider a regime in which the seller is free to gather the receipts, but must disclose them to the buyer in the event that they are gathered. In considering this last regime, we assume that a court is able to determine whether the seller in fact gathered the information and that the sanctions of the law are sufficient such that, if the law is in effect, rational sellers will comply with it.

To solve the game under a regime of voluntary disclosure, we need to identify the possible equilibria. Let us posit that there is an equilibrium in which all sellers become informed. To be a perfect Bayesian equilibrium, the buyers must believe that all sellers are informed. For this reason, the unraveling result will hold and there will be complete revelation of information. All owners of houses that do not need new furnaces will reveal this information and receive $200. All owners of houses that need new furnaces will remain silent and receive $180. Before learning about the furnace, the seller expects to receive $190 for the house.[23]

In this equilibrium, all sellers become informed, and buyers believe that their sellers are informed. If a seller deviates and does not learn about the furnace, buyers, given the beliefs that they hold in the equilibrium, will infer from silence that the seller has learned that the house needs a new furnace and is remaining silent. Hence, they will pay the seller who deviates only $180. Thus, in this equilibrium, the value of information to a seller is $10, the difference between what the seller can expect to receive by becoming informed (and having the chance to reveal favorable information) and what the seller can expect to receive by not becoming informed (and having to remain silent in all events). Hence, this equilibrium can exist only if all the sellers can gather the information for less than $10. (If any could not gather the information for this amount, they would not become informed and this equilibrium could not exist.)

Let us ask if an equilibrium can exist in which only some of the sellers become informed. Is it possible, for example, for there to be a distribution of costs among sellers such that only half (those with low costs of information acquisition) become informed? In such an equilibrium, buyers would not be able to determine whether a seller who remains silent needs a new furnace. Silence might reflect an absence of knowledge—the seller did not acquire the information—or it might reflect a desire to hide the bad news—the seller learned that the furnace should be replaced. Nevertheless, the action of the seller would lead the buyer to update the belief that a seller was equally likely to have a good furnace as a bad one.

In such an equilibrium, ¼ of the sellers both have good furnaces and reveal the information. (Half gather the information, and ½ of those will have good furnaces.) The remaining ¾ remain silent. One-third of these silent sellers are the informed sellers who found out that their furnace needed replacing; another ⅓ are the uninformed sellers with furnaces that need replacing; and the final ⅓ are the uninformed sellers with good furnaces.

The silence of a particular seller conveys information that should cause rational buyers to alter their original beliefs. Two-thirds of all silent sellers have bad furnaces. Bayes's rule therefore requires that a buyer who observes that a seller is silent change from believing that the seller's house is as likely to have a bad furnace as not to believing that there is a ⅔ chance that the furnace is bad.

The buyer will, as a result, be willing to pay the silent sellers an amount that reflects the ⅔ chance that the value of the house is $180 and the ⅓ chance that it is worth $200, or $186.67.[24] In this equilibrium, the informed seller expects to receive $200 or $186.67 with equal probability, before taking into account the costs of gathering the information. A seller therefore expects to receive $193.33 from gathering information, less the costs of gathering it. Because the difference between $193.33 and $186.67 is $6.67, this equilibrium can exist if only ½ of the sellers are able to gather the information for less than this amount and the other ½ are not.

The last type of equilibrium to consider is the one in which none of the sellers becomes informed. If, in equilibrium, no sellers become informed, a silent seller expects to receive $190. This equilibrium is sustainable only if no seller can do better by becoming informed and disclosing the information if it is favorable. If a seller were to gather information, there would be a 50–50 chance of the seller's receiving $200 instead of $190. A seller, in other words, would expect to realize

$195 from deviating before taking into account the costs of gathering the information. As long as the costs of gathering the information exceed $5, the seller's best response given the actions and beliefs of the buyer is to remain uninformed. Hence, this equilibrium can exist if the costs to all sellers exceed this amount.

At this point, we can see that the solution to the game turns on the costs to the sellers who can gather information cheaply relative to those who cannot. If, for example, all sellers can gather the information for less than $5, the only equilibrium is one in which all gather the information. Given the costs of gathering information, the combination of actions and beliefs in which some sellers do not become informed cannot be an equilibrium. The benefit from deviating from such an equilibrium for an uninformed seller (a 50–50 chance of getting $200 instead of $190, or $5) exceeds the costs of gathering the information. Hence, this proposed equilibrium must be rejected because it is one in which the actions of some sellers are not a best response given their beliefs and the actions and beliefs of others.

In short, the solution to the game turns on the costs that the sellers face. In some of these cases, there are two possible equilibria: one in which all the sellers become informed, and the other in which none does. Whether we can eliminate one or the other as a solution to the game depends on whether we can refine the perfect Bayesian equilibrium solution concept. The major lesson of this model, however, is that solutions to this game do exist in which parties spend resources gathering information even though the information brings no social benefits.

The game in which sellers must disclose any information they gather is easier to solve. Sellers who gather no information receive $190. Sellers who gather information have as good a chance of receiving $200 as $180. They too can expect to receive $190, but from this amount they must subtract their costs of gathering information. A player has a lower payoff from gathering information than not and therefore will not do so. No player ever gathers any information. Because the information has no social benefit, the outcome under this legal regime is an improvement over the earlier one—assuming, of course, that courts have some means of determining whether a seller has in fact gathered the relevant information. (If a court has no means of making such a determination, then there is no way of enforcing a law that requires disclosure.) A law that forbids disclosure would, in theory, have the same effect.

A legal rule that required the seller both to gather and then to disclose this kind of information would not make sense in this context.

The game in which all sellers gather the information about the furnace and disclose it generates the worst possible outcome because, if the sale is to go forward, the seller must gather the information, no matter what it costs. Because the information has no value, all the resources spent on gathering the information are wasted.

As we have already seen in the case of *Laidlaw v. Organ,* situations can also easily arise in which the buyers have information and the sellers do not, but the information still has no social value. We can see another such case by making a small change in the example with the buyer and the seller. The buyer might be in a position to find out whether a new highway was being built nearby that increased the value of the seller's land, but did not change the way it should be used in the interim. The outcome under different legal regimes is much the same. If buyers were forced to disclose, they would never gather the information. If they were forced to gather and disclose the information, they would either waste resources gathering and disclosing it or they would not buy the land at all. Under a regime of voluntary disclosure, we could also have buyers gather and disclose the information, except that they would disclose only information that lowered the price, rather than information that raised it.

Let us change the previous example and assume that the information is not about whether the house needs a new furnace, but rather about how the house was constructed and how it has been maintained over the years. As before, this information reveals how much a house is worth. A house that has asbestos or lead paint, for example, is worth less than one that does not. This information—unlike that in the case of the furnace—also has additional value. An owner that has this information knows, for example, what repairs need to be done.

For simplicity, let us assume that a house is either well-built and worth $200 to its owner, or badly built and worth only $180. Information about the house tells the seller (and the buyer if the seller discloses it) how well the house was built. In addition, the information, if disclosed, increases the value of the house to a new owner by $5, regardless of how well it was built. The information about how the house was constructed is valuable because it provides the owner with information about what repairs to make and when to make them. Half the sellers can assemble the information for $4, but it costs the other sellers $8 to gather it. Because the information brings a benefit of only $5, the efficient outcome is for low-cost sellers to gather the information, but not for high-cost sellers to do so.

Let us first explore the solution to this game under a voluntary disclosure regime. In one equilibrium, all sellers gather the information and disclose it, while the buyer believes that any seller who remains silent has gathered the information and learned that the house is badly built. A seller who discovers that the house is well built reveals this and receives $205, less the $4 or $8 cost of gathering the information. A seller who discovers that the house is badly built discloses it and receives $185, again less the costs of gathering the information. In both cases, sellers receive the value of the house ($180 or $200) and $5 for the information.

In this equilibrium, a seller enjoys an expected payoff of $195, less the cost of information (a payoff of $191 or $187, depending on the seller's type). Gathering the information and disclosing it is a best response for both types of sellers in this proposed equilibrium. Given the buyer's beliefs, the buyer would infer that any seller who refuses to disclose the information has learned that the house is badly built. Such a buyer would therefore offer only $180 to a silent seller because without the information itself the house is worth only $180. This amount is less than the expected payoff to either type of seller. The key to this equilibrium is that, given the strategies of the sellers, no seller remains silent, and thus nothing constrains the buyer's beliefs about the sellers who remain silent.

We now must ask whether there are other combinations of strategies and beliefs that are plausible solutions to this game. Consider, for example, any equilibrium in which no sellers gather information. Such a proposed equilibrium gives sellers an expected payoff of $190. Half the sellers, however, are not choosing a best response by remaining uninformed. Those whose costs of gathering information were only $4 could gather the information, receiving a payoff of $201 in those cases in which the news was good, and a payoff of $186 in those in which it was bad. (The sellers with bad houses would remain silent and still receive $190, but they would have spent $4 on information.) The expected payoff to the low-cost sellers from deviating is thus the average of $201 and $186, or $193.50. Because the low-cost sellers are not choosing a best response in this proposed equilibrium in which all sellers are silent and receive $190, we must reject this solution to the game.

We next ask whether a perfect Bayesian equilibrium can exist in which only the low-cost sellers become informed. This combination of actions and beliefs forms an equilibrium only if it is in the interest of high-cost sellers to remain uninformed at the same time that the buyer believes that they remain uninformed. The buyer can have this belief only if remaining uninformed is a best response for the high-cost sellers. To test whether it is, we have to ask what payoff high-cost sellers would enjoy if they gathered information but all other actions and beliefs remained the same—including the buyer's belief that high-cost sellers remain uninformed. We then compare this amount with the payoff that high-cost sellers receive when they remain uninformed.

High-cost sellers receive $186.67 when they remain uninformed and the buyer believes that they are uninformed.[25] We have to ask if high-cost sellers can do better by taking another action, given that the buyer has this belief. If high-cost sellers spend $8 on information and discover that their house is well built, they net $197 ($205, less its costs of $8). If high-cost sellers learn that their house is badly built, they remain silent, and, because the buyer believes that they are uninformed, they continue to receive $186.67, less the $8 they spent on the information, or $178.67. (High-cost sellers with badly built houses remain silent when the buyer believes that they are uninformed. They receive only $185 when they disclose the information.) High-cost sellers who become informed therefore enjoy an expected payoff of $187.84 (the average of $197 and $178.67). This amount exceeds the payoff of $186.67 that high-cost sellers enjoy when they remain uninformed and the buyer believes that they are uninformed.

We can now reject the proposed equilibrium in which high-cost sellers remain uninformed. High-cost sellers prefer a payoff of $187.84 to one of $186.67. We should reject a solution to the game that requires that the high-cost sellers remain uninformed when they could do better by becoming informed. In any solution to the game, players should take the actions that they believe lead to the highest payoffs, given the actions and beliefs of the other players.

Finally, we can eliminate the equilibrium in which all sellers gather the information but only those who find that their houses are well built disclose it. Such an equilibrium can exist only if buyers are willing to pay those sellers who remain silent at least the $185 the buyer would pay if those who gathered the information and received bad news disclosed it. Buyers, however, must have beliefs that are consistent with the actions of the sellers, and, in this proposed equilibrium, any seller who remains silent has a badly built house. Buyers are willing to pay such sellers only $180. (Without information about how to repair the house, a badly built house is worth only $180, not $185.) This amount is less than the $185 that the sellers of badly built houses would receive by disclosing the information. Hence, they are not making a best response in this equilibrium.

By exhausting all the alternatives, we have established that the only perfect Bayesian equilibrium is the combination of strategies and be-liefs in which all the sellers gather and disclose information and the buyer believes that any seller who does not disclose information has a badly built house. This outcome is undesirable because half the sellers are spending $8 to gather information that brings a social benefit of only $5.

There is a different outcome under a legal regime in which sellers who know whether their houses are well built must reveal this infor-mation, but have no obligation to gather this information in the first place. If there is mandatory disclosure, an informed seller will receive $205 or $185 with equal probability, or an expected return of $195, less the expenses of gathering the information. An uninformed seller will receive $190 and face no expenses. Low-cost sellers will gather the in-formation and enjoy an expected net return of $191.[26] High-cost sellers, by contrast, are better off remaining silent and enjoying a return of $190 rather than spending $8 gathering information and enjoying an expected net return of only $187.[27]

When sellers must disclose whether their houses are well or badly built, they can increase the amount they receive from their buyers by gathering the information only to the extent that the information has value independent of the way it affects the price. A rule that requires disclosure of socially valuable information sometimes discourages in-dividuals from gathering information in the first place. The rule does not have that effect in this case. Once disclosed, the information has value to the buyer only if the buyer purchases the house. Because the seller owns the house, the seller also "owns" the information and there-fore is able to capture its value even after disclosing it.

The last legal regime to consider is the one in which all sellers are required to gather the information and disclose it. Here we face the same undesirable outcome that we saw in the case in which disclosure was voluntary. This rule requires sellers to gather information even when gathering the information is extremely costly. This rule never leads, however, to an equilibrium in which a seller gathers the informa-tion but does not disclose it. Such equilibria are possible under a volun-tary disclosure regime. In principle, cases can arise in which we are better off in a regime in which sellers must gather information and disclose it. The losses from forcing parties to gather information when they should not may be smaller than the costs that parties incur gather-ing information that they then do not disclose when it proves unfavor-able.

Buyers are sometimes better positioned to gather information. In many cases, however, buyers can acquire the information after they buy an asset. At this point they will have the right incentive to gather information because they incur all the costs and enjoy all the benefits of that information. We should, however, consider the case in which information gathering, for some reason, must occur before the sale and in which only buyers are able to gather it. A voluntary disclosure regime may be inefficient when information has no social value because equilibria can exist in which buyers gather information. By contrast, no buyer will gather the information in a mandatory disclosure regime. In such a regime, the seller can extract the benefit of the information by raising the price once the information is revealed. The information has value only to the person who owns the asset. Because the party with the information does not own the house, this case is different from the one in which the seller had the information and was required to disclose it. Requiring a person with information about an asset to disclose that information to the asset's owner has a different effect from requiring the owner of an asset to disclose information about it to someone else. The transfer of information itself makes the recipient better off in the first case, but not in the second.

In some cases, the transfer of information before sale is useful because it ensures that buyers and sellers are matched more effectively. We can illustrate with a variation on the previous example. Let us posit that there are two groups of buyers of equal size. There are those who do not mind having a badly built house as much as others. To be sure, such a house requires more repair, but it also provides them with a chance to pick out paint colors, appliances, and so forth that suit their tastes. The other buyers do not care about customizing the house and receive no benefit from making repairs. The first group values a badly built house at $190 and the second values it at $170. Both would prefer to live in a well-built house, which they again value at $200. Let us continue to assume that there are a large number of houses on the market and that half are well built and half are not. Sellers can acquire verifiable information about their houses at a cost of $8 or $4, as in the model above. If a seller does not reveal information, however, a prospective buyer can, at some cost, gather the information and learn whether the house is well built.

In a voluntary disclosure regime, half the sellers acquire information and half do not. A seller who discovers that a house is badly built remains silent. Only ¼ of the sellers will announce that they have well-built houses. Thus, some buyers who put the lowest value on owning a badly built house must try to buy a house from a silent seller. These buyers find it in their interest to investigate. If they discover that the house is badly built, they could reveal it and offer to buy the house for $170. The seller, however, would not accept the offer because there are others in the market who value the house at $190.

Eventually, all the badly built houses will be sold to buyers who put the highest value on badly built houses and all the well-built houses will go to the others. In the process, however, too much money will be spent gathering information. Some buyers who put the lowest value on badly built houses will find sellers who reveal that they have good houses. Most of them, however, will have to spend resources gathering information about houses until they find one that is well built. These buyers will typically have to investigate three houses.[28] Some houses will be investigated more than once.[29] A mandatory acquisition and disclosure law for sellers can do better if the cost of gathering the information for the high-cost sellers is not too high. Such a law works by reducing or eliminating the redundant acquisition of information.

In this section, we have examined the effects of disclosure laws on a particular problem involving verifiable information. One party has information or may be able to acquire information that the other does not have and cannot acquire. The first party does have the ability to convey the information to the other once it is acquired, but may not find it advantageous to do so. The information may or may not have social value. A court is able to determine after the fact what information the first party has acquired, but the other party cannot. This problem provides a sense of how we should think about laws governing the disclosure of information and forces us to focus both on the social value of information and on the ability of the parties and the courts to learn about it.

The perfect Bayesian equilibrium solution concept we use to analyze this problem rests on stronger assumptions than the ones we developed in earlier chapters. Although a player might not know whether the other possessed the relevant information, each player knows everything else about the structure of the game, including the prior beliefs that the other player holds. In addition, each player knows that the other player knows the game as well, knows that the other player knows that that player knows it, and so forth ad infinitum.

We return to the problem of *common knowledge*—that all the details of the game are known and are known to be known—when we examine the problem of predation in Chapter 5. At this point, however, it is worth noting that we cannot justify the assumption of common knowledge by asserting that it is consistent with the way people generally behave. Many actions that we see, such as speculative trading in commodities markets, are inconsistent with the assumption of common knowledge.[30] The justification for making this assumption rests rather on the idea that this tool, like others in game theory and economics more generally, sheds light on the dynamics of the interaction between parties. In this case, the solution concept allows us to focus rigorously on the kinds of inferences that can be drawn about information from the actions that parties take.

Observable Information, Norms, and the Problem of Renegotiation

Up to this point in this chapter, we have assumed that information is known to only one of the parties. The question we have been facing has been whether the informed party will disclose the information and, if legal rules mandate disclosure, whether that party will bother to become informed in the first instance. In the rest of this chapter, we want to shift our focus to the problem that arises when the information is known to both parties, but not to any third parties. Information in this case is said to be *observable but not verifiable*. In these cases, there are facts about the transaction between the parties that, although known to the parties themselves, cannot be communicated to a court. Hence, these facts cannot be the basis on which a court decides any dispute that is brought before it. The principal problem that the parties face in this context is creating a mechanism that ensures that both behave optimally even though courts know less than the parties themselves. These problems typically arise in contract cases.

We begin by considering an aspect of one of the best-known cases involving contract renegotiation. During the Alaskan fishing season of 1900, sailors under contract to a salmon packer demanded a pay raise after it was too late to replace them. The sailors claimed that the packing company that hired them provided nets of poor quality. Because their pay was determined by the size of the catch, the sailors asserted that they were entitled to additional pay. The packer acceded to their demands, but then reneged once the season was over and the sailors returned to San Francisco. The sailors took the packer to court, but their efforts to force the packer to pay them proved unsuccessful. The case they brought, *Alaska Packers' Association v. Domenico*,[31] has become one of the leading cases on the problem of "economic duress" in contract law, the problem that arises when one party demands changes in the original agreement after the other has already committed itself to performing.

An important factor that may have contributed to the court's finding against the sailors in *Alaska Packers* was the inadequacy of contract damages. If parties cannot write a contract in which the damages that one party owes the other in the event of breach are sufficiently large, that party may threaten to breach before the contract has run its course. The other party may have committed so many resources to the contract and the remedy under the contract may be so weak that the other party is better off renegotiating the contract than suing for breach. We want, however, to focus on the information problem in *Alaska Packers*—specifically, the inability of the court to determine whether the nets the packer gave the sailors were defective.

The packer and the sailors both benefited from having high-quality nets. This is itself the source of the problem. As long as one party or the other is responsible for the condition of the nets, that party will enjoy only part of the benefit of having high-quality nets, but will incur all the costs. Hence, that party will have an incentive to spend too little on the nets. As soon as it makes sense to have the sailors' pay dependent on the size of the catch, the parties have to seek a way to force the party who buys the nets to take account of the benefit of high-quality nets to both parties.

If a court could verify the condition of the nets, the contract could have easily been written in a way that protected the sailors. Unlike the sailors, the packer is unlikely to be judgment proof, and the sailor's ability to bring an action for damages against the packer would ensure that the packer had the incentive to provide good nets. The court, however, could not do this. The packer and the sailors may well be able to observe the quality of the nets and the condition of the boats, but it could be very difficult and expensive to have a court make informed judgments about such an issue. The parties therefore could not write a clause in their contract that was tied explicitly to the condition of the nets. Clauses in a contract must be ones that a court can enforce.

When information about the condition of the nets is not verifiable, an equilibrium can emerge in which the sailors are paid a wage based on the expectation that the packer will provide bad nets and the packer does in fact provide bad nets. The packer's promise to provide good nets is not credible, and hence the sailors will infer that the packer will not provide them. A legal rule cannot mandate that the packer provide good nets, because, by assumption, information about the nets is not verifiable. There is no way a court could ever learn whether the packer lived up to its obligation.

When parties confront a situation in which observable but nonverifiable information looms large, they can take steps to ameliorate the problem. The sailors might, for example, insist that the packer take an action that, once carried out, gives the packer an incentive to ensure that the nets are in good repair. The packer, for instance, might commit in the contract to spending a certain amount of money on nets, and this information might be verifiable. Once the packer must spend money on nets, it might be in the packer's interest to provide good ones.

The sailors might instead insist on an alternative to the courts as a dispute resolution mechanism. For example, they might demand the right to insist on arbitration. If they find the nets unsatisfactory, the sailors can begin a procedure in Alaska in which they can pick an arbitrator, the packer can pick an arbitrator, and the two of them can pick a third. The three together can rule on the condition of the nets. The condition of the nets might be nonverifiable information to the court, and yet be information that is verifiable to the arbitrators. The only issue for the court would be whether the parties abided by the decision of the arbitrators. One needs to ensure that the law makes it possible for parties to adopt mechanisms, such as alternative dispute resolution, as ways of avoiding such problems.

Some contractual relationships, particularly long-term contractual relationships, resist easy solution. When parties enter into a long-term contractual relationship, each party typically gathers enormous amounts of information about the other, such as the demand for a product, the quality of the design, the productivity of a worker, the quality or quantity of promotional service provided by a distributor, or the match between the employees of two businesses in a joint venture. Any of these might be hard to summarize in a statistic that can be the basis for a contractual term that a court could enforce.

Moreover, the information may be observable but nonverifiable because those who have invested enormous resources in the relationship—the two parties to the contract—may be readily able to tell whether the other is performing as promised, but this information may be unavailable to third parties. These problems are magnified if conditions are subject to change over time. We have focused on problems that are most likely to arise in business contexts, but the problem may be even more acute elsewhere. Consider, for example, the problem that arises in litigation between spouses over the custody of a child. It may be common knowledge between the spouses that one cares little about the child. If the other spouse, however, has no way of persuading the court that this is the case, the first spouse may be able to use the custody issue as a means of extracting a larger property settlement.

Problems involving observable but nonverifiable information are likely to arise in the case of long-term contracts in which conditions change over time. The parties themselves may know how the conditions differ from what was expected, but a court may not. Renegotiations must take place in this climate, a fact that the parties should anticipate at the time they enter into their contracts. In the next section, we examine the implications for contract law when confronting this problem.

Optimal Incentives and the Need for Renegotiation

Contracts that last for a long time, such as a contract between a coal mine and a power plant built next to it, are often written in a way that explicitly contemplates renegotiations when economic conditions change.[32] The contracts are *incomplete* in the sense that the obligations spelled out in the contract do not lead the parties to realize all potential gains from trade. Because courts are limited in the information they can acquire, the contracts that parties write may fail to take sufficient account of all contingencies.[33] These contracts, however, typically provide that neither party has a legal duty to renegotiate. Neither party can sue the other if renegotiations prove unsuccessful. It is necessary to explain the apparent incongruity between a contract that contemplates renegotiation and one that imposes no legal duty on any party to renegotiate.

The existence of observable but nonverifiable information can affect the ability of parties to renegotiate their original agreement. This in turn can affect the way the original agreement is written. After the fact, the terms of the contract may not fit the present needs of the party, but this alone does not make it desirable for a court to reshape the terms of a contract. When the parties anticipate that a contract will be renegotiated, they may want to be able to create a set of terms that would be a good starting place for the negotiations. The way the terms shape the course of renegotiation rather than their suitability to the conditions in which the parties find themselves may be what matters most. Courts must be willing to enforce these terms if the parties are to be able to create the appropriate climate for renegotiation in their original contract.

In this section, we develop a model that examines this problem. This model suggests that we may not want legal rules that require courts to call off obligations or modify the terms of the contract when such terms seem out of step with current conditions. The parties, before the fact, might prefer a fully enforceable, unambiguous, albeit incomplete, contract that did not track their needs at some future time to one that was more flexible. The former might well create a better environment for renegotiation which, given that the contract is necessarily incomplete, would have to take place in any event.

Consider a contract between the supplier of an input, "Seller," and a downstream manufacturer, "Buyer." The parties enter into a long-term contract because Seller must decide the quality of inputs to supply. The more money Seller invests in adapting its plant to produce to Buyer's specifications, the higher the quality of the product. In addition, Buyer must retool its manufacturing plant to accommodate the characteristics of Seller's product.

Each party's investment, although observed by the other, is not verifiable, so a contract cannot be made contingent on these decisions. In addition, the value of the product to Buyer is subject to significant uncertainty, so that Buyer's willingness to pay for Seller's product will vary. For simplicity we assume that Buyer purchases either one unit or zero units from Seller. Buyer and Seller can observe Buyer's demand after the contingencies resolve themselves, but a court cannot.

We wish to formalize the idea that information that is unknown at the outset becomes known during the course of the contract. In most contracts, this information emerges only gradually over time. We can capture the essence of the problem, however, in a model in which events proceed in discrete steps. In this model, there are five successive periods. In the first, the contract is signed. In the second, the buyer and the seller both make their investment decisions. In the third period, the uncertainty is resolved. In the fourth period, the parties then have a chance to renegotiate, and, in the fifth, the parties decide whether to carry out the terms of the contract or breach it.

In this model, the value to Buyer of a unit of the input is the sum of three elements, $v + k_s + k_b$. The element v is the part of the value of a unit to Buyer that is uncertain. The value of v ranges from 0 to 1 and is equally likely to be found at 0 or 1 or anywhere between the two. The element k_s represents Seller's quality investment, and k_b represents Buyer's investment. Seller's investment costs $2k_s^2$, and Buyer's investment costs k_b^2.[34] Seller's marginal cost of production is assumed to be 0. When we solve for the efficient level of investment, we discover that the optimal levels of k_b and k_s are ½ and ¼, respectively. The expected social surplus is ⅞.[35]

In this model, the efficient level of investment does not occur when there is no contract. Seller invests too little in making a quality product and Buyer spends too little retooling its plant to take advantage of

Seller's product. Because there is no contract, the parties bargain over a price after investments have been made. There is no private information, because each can observe the investment decision of the other. Buyer values the good at $v + k_s + k_b$. This value exceeds the marginal cost of making this product, which, for simplicity, we have assumed to be 0. Because there are no transaction costs at this stage, the parties should reach an agreement. In the absence of any other information about the parties, we assume that they divide the potential gains from trade evenly between themselves.[36] The price is therefore $\frac{1}{2} \times (v + k_s + k_b)$, and each party enjoys only $\frac{1}{2}$ of the social value of its investment. The other party appropriates the other $\frac{1}{2}$. For this reason, each party invests too little at the outset. The equilibrium levels of k_b and k_s are $\frac{1}{4}$ and $\frac{1}{8}$, respectively. The expected surplus is $\frac{25}{32}$, which is less than the first-best surplus of $\frac{7}{8}$.

Even when parties can enter into a contract with each other, the existence of nonverifiable information makes it impossible to give each party the incentive to make the socially optimal level of investment. For each to have the correct incentive, we would have to make each party the effective residual claimant for its investment decision—and this is not possible. If we make Buyer the residual claimant, Seller will have no incentive to invest, and vice versa. We must make a trade-off between ensuring that Seller has the right set of incentives and ensuring that Buyer does also. Nonetheless, Buyer and Seller could improve upon the outcome that exists when there is no contract by entering into a long-term contract with each other.

Given that other information is nonverifiable, the contract between the parties must focus exclusively on price and quantity. Under our assumptions, Buyer will need either one unit or none. A contract that fixes quantity at one and also fixes price does not create incentives for Seller to invest. Seller is assured of getting the fixed price independent of its investment decision. A fixed quantity, variable-price contract will not give both sides an incentive to invest either. If Seller gets to choose a price that Buyer can accept or reject, Seller will set the price equal to the value that Buyer places on it. This contract leads to underinvestment on the part of Buyer. Buyer enjoys no ex post surplus no matter what investment it makes, so it makes none. Similarly, if Buyer gets to choose a price, it will set the price equal to marginal cost, which is zero, and Seller has no incentive to invest.

The parties, however, could enter into a fixed price, variable quantity contract in which Buyer can choose either to buy one unit at a fixed price or not buy any at all and pay no damages. When uncertainty is resolved, the product may be worth more than Buyer is required to pay for it. In this event, Buyer would exercise its right to buy. Because all uncertainty has been resolved, the surplus that arises from the trade is now fixed. Buyer can force Seller to sell the good at the fixed price. Hence, Seller has no way to force Buyer to renegotiate and pay a higher price. Because Seller has nothing to gain from renegotiating with Buyer, Seller can credibly choose a strategy in which it does not enter into any renegotiations. Once Seller adopts such a strategy, Buyer cannot credibly threaten to refuse to exercise its right to buy the product.

If uncertainty is resolved in a way that the value of the product is less than the price in the contract, Buyer values the product less than the contract price. In this case, Buyer's threat to walk away is credible because it is not in Buyer's interest to buy the product unless the price is lowered. Seller will be willing to enter into renegotiations because the marginal cost of producing the product is $0 and there are potential gains from trade. We must make an assumption about the price on which they agree in this renegotiation game, but the qualitative results of the model do not depend on this assumption.[37] To keep things simple, let us assume that Seller will lower the price only to the point at which Buyer is indifferent between buying and not.

Each party has some incentive to invest under this type of contract. Buyer receives the surplus when the value it places on the product after the uncertainty is resolved exceeds the contract price. As Buyer increases its investment, this surplus increases. The lower the price in the contract, the more Buyer will enjoy this surplus, and hence the more incentive it has to invest.

If the value that Buyer places on the product turns out to be low relative to the price set out in the contract, there will be renegotiation. Seller will receive only the value of the product to Buyer. Because the value of the product increases as Seller's investment in product quality increases, Seller's expected return increases as it invests in quality until it reaches the optimal level of investment. The greater the contract price, the more likely Seller will be to capture the benefits of increasing the quality of the product.

In the contract, Buyer and Seller will set the price at a level that trades off these incentives optimally. If this price gives too much surplus to one side relative to its initial bargaining power, there can be a transfer at the time the contract is signed. In this way, we can ignore issues of ex ante bargaining power. When we solve this problem, we discover that the optimal price is $3/4$, $k_b = 1/3$ and $k_s = 1/12$.[38] The social surplus is $19/24$. This amount is less than the first-best, but greater than $25/32$, the surplus in the absence of a contract.

In equilibrium, the contract price exceeds Buyer's value $2/3$ of the time. Buyer is given a greater incentive to invest in the contract, because Buyer enjoys greater returns from investing under our assumptions than Seller does. In the absence of a contract, the surplus is split in a way that is not related to the relative abilities of the parties to invest in the contract. This model shows how contracts may play an important role in structuring bargaining power in renegotiation, even in a case such as this, in which negotiations take place only when no trade at all takes place under the contract.

The court in *Aluminum Co. of America v. Essex Group, Inc.,* required parties to reform the price adjustment mechanism in their contract that failed to track market fluctuations in the price of aluminum in the wake of energy price changes in the early 1970s.[39] The model we have developed in this section, however, suggests a reason courts should be reluctant to impose changes on such long-term contracts. The contract might not have been the one the parties would have written if they had known how the cost of producing aluminum at a particular plant and the demand for the product would have changed the way they did. This fact alone, however, may not justify departing from the price escalation clause that the parties adopted in their contract. The parties may never have intended the simple price escalation clause they adopted to track changing conditions perfectly. Rather, they may have recognized that it would set the climate for renegotiations should they become necessary.

Other provisions that we often see in long-term contracts, such as rights to terminate, may not be clauses that parties plan to exercise; it may nevertheless be important for courts to respect them because they may be needed to ensure that the renegotiations take place that advance the parties' interests. As we have seen elsewhere, the solution to a game often depends crucially on the payoffs of strategy combinations that are off the equilibrium path. A court that tries to impose conditions that it thinks are in the long-term interests of the parties may do more harm than good because, by modifying these payoffs, the court may inadvertently change the equilibrium outcome of the game.[40]

Limiting the Ability of Parties to Renegotiate

In some cases, renegotiation affects only the relative wealth of the parties. There is no social gain from these renegotiations. In other instances, as we showed in the last section, renegotiations may make the parties better off. A third possibility also exists: Renegotiations make sense at the time of the renegotiations, but the parties would nevertheless want to prevent them ahead of time if they could. In some cases, it might be in the interest of the parties to prevent renegotiations.

Consider a contract between a defense contractor and the government to develop and build a new fighter plane. The government wants the contractor to develop the plane in a reasonable period of time and at a reasonable cost. If the government cannot observe the efforts of the contractor, there is a moral hazard problem. If the contract gave the contractor a fixed sum of money, the contractor would not take sufficient effort. This is an example of the classic principal-agent problem. In many cases, the optimal contract will give the government the right to cancel the project if delays are too long or cost overruns are too great. Such a contract, however, will be subject to a significant renegotiation problem. Imagine that there are large delays and cost overruns. It may not be optimal to cancel the project at this point, given that an enormous amount of resources has already been sunk into the project. The incremental costs of continuing may be small relative to the incremental gains. There are gains from renegotiating the original contract; both parties can be made better off by ripping up the old contract and signing a new one that includes continuation.

In this case, renegotiations are not desirable before the fact, even though they are desirable after the fact. The possibility of renegotiations undermines the contractor's incentives. The contractor will work less hard than it should because, even if the government has the right to cancel the contract, it will never be in its interest to do so. Instead, the government will renegotiate. An incentive contract that can be renegotiated cannot work as well as one that cannot be renegotiated. Writing such a contract, however, is difficult. Assume that there is a contract that cannot be renegotiated and that it provides the contractor with the incentive to take the optimal amount of effort. What happens in the event that there are still delays and overruns?

In this case, the government knows that, because the contractor has the right incentives, the delays and overruns are most likely due to bad luck rather than to insufficient effort. Canceling the contract at this point would serve no purpose. The government gains nothing from canceling and the contractor has done nothing that warrants punishment. In such a situation, the incentive to renegotiate is enormous. Nevertheless, the commitment not to renegotiate created the optimal incentives in the first place. As long as the government can renegotiate, it cannot be sure that the contractor acted optimally.

There is no simple way around this inability to commit to no renegotiation. Our entire legal system would have to change dramatically in order to use legal rules to make contracts nonnegotiable. It is insufficient to write a clause into a contract that says, "We promise not to renegotiate this contract," with enormous penalties if the clause is breached. If it turns out to be in the parties' interest to renegotiate, all they have to do is agree to rip up the old contract and write a new one stating that the clause in the old contract is void. Parties will be able to circumvent any legal limits on renegotiability if they can rewrite the new contract in such a way that neither ever has the incentive to object to it.

When a commitment not to renegotiate is extremely valuable, contracting parties might try to use a third party to enforce their agreement. They could write a contract that says, "If we renegotiate, we each agree to pay a third party one billion dollars." Such a contract, however, may not work either. Let us assume that parties write such a contract, but a situation arises in which renegotiation would, after the fact, make both parties better off to the extent of $1,000. The parties could go to the third party and say, "We will not renegotiate our contract because it costs us $2 billion and saves us only $1,000. Therefore, you will get nothing. If you agree to rip up our contract with you and release us from our obligations to pay $2 billion, we shall give you $100."

The third party prefers $100 to nothing, so that contract is renegotiated as well. The third party may be in this business and may wish to protect a reputation for not renegotiating. If this is the case, the private system may be able to do some good. Alternatively, the players might oblige themselves to many diverse third parties in the event of a renegotiation. The high transaction costs involved in reaching an agreement with the diverse parties may provide the deterrent that ensures that renegotiations do not take place.

Summary

Legal rules that affect the disclosure of information turn crucially on the kind of information involved. In this chapter, we began with cases in which a player possesses information that can be readily disclosed and in which the other player knows that that player possesses such information. In these situations, legal analysts must take account of the unraveling principle. Not only may legal rules be unnecessary to ensure that information is disclosed when disclosure is desirable, but there may be significant obstacles to implementing legal rules designed to prevent the disclosure of information.

More complicated cases arise when information can be readily disclosed, but the other player has no way to distinguish the player who possesses information and remains silent from the player who does not possess the information in the first place. In this context, there may be room for legal rules mandating the disclosure of information, but such rules are not always desirable and they can work only if the court is able to make a distinction between silent but knowledgeable players and ignorant ones.

We also showed in this chapter why legal analysts must take special note of those cases in which the information is observable to the parties but not to the court. In particular, we developed a model that shows how the parties might write their contract so that the renegotiations that would have to take place when circumstances changed could be structured so that both parties had the incentive to act in a way that was in their mutual interest before the fact.

In the next chapter, we confront a radically different problem that the legal analyst must take into account—those cases in which a player possesses information but has no way of conveying this information to the other player, and a court has no way of discovering it after the fact. The problem in these cases is one of determining whether players can draw inferences from the actions that others take and of understanding how legal rules can affect, for better and for worse, the ability of players to draw such inferences.

Bibliographic Notes

The economics of information.   Good textbooks on information economics include Hirshleifer and Riley (1992) and Laffont (1990). Information plays an important role in modern economic theory, and there are extensive discussions in most advanced textbooks, including Varian (1992) and Kreps (1990b).

Perfect Bayesian equilibrium.   The concept of perfect Bayesian equilibrium is closely related to sequential equilibrium, first analyzed by Kreps and Wilson (1982). For a discussion of the differences, see Fudenberg and Tirole (1991b). Gibbons (1992) contains numerous examples of solving for perfect Bayesian equilibria.

Verifiable information and unraveling.   The unraveling result is analyzed in Grossman (1981), Grossman and Hart (1980), and Milgrom (1981). Jovanovic (1982) considers the case where revelation is costly and shows that there is socially excessive revelation. Okuno-Fujiwara, Postlewaite, and Suzumura (1990) is the most complete theoretical treatment of conditions under which unraveling does or does not hold. Easterbrook and Fischel (1991) discusses the problem of unraveling in the context of securities regulation.

Shavell (1989) and Shavell (1991) analyze cases of disclosure in which uncertainty exists about whether the silent party is informed. Our model of disclosure develops this model by focusing specifically on the question of the relationship between legal rules and incentives both to acquire and to disclose information. Moreover, we identify the need to confront the unraveling problem and the question of whether a court will have access to information that the uninformed party does not have at the time of the transaction. Our typology of legal rules also includes rules that require a party both to gather and to disclose information, in addition to legal rules that require disclosing the information that parties already possess. We also show that problems that arise when buyers are the ones who may acquire information matter most when there are different types of buyers and sellers and we want them to be matched in the optimal way.

Modeling the disclosure of information. Kronman (1978) contains an early discussion of the incentives to obtain costly information prior to trade without focusing on disclosure rules. Fishman and Hagerty (1990) addresses the question of what a party should reveal and the effects of mandatory disclosure when the amount of information revelation is constrained. Geanakoplos (1992) explores the idea of common knowledge and the nonspeculation theorem, which shows that common knowledge of rationality and of optimization eliminates speculation.

Renegotiation. The literature on renegotiation and long-term contracts has its root in the literature on the theory of the firm that focuses on the differences between organizing activities through ownership, markets, or contracts. The pioneering paper is Coase (1937). Many of Oliver Williamson's extensive writings deal with these issues; especially useful is Williamson (1985). Another important paper is Klein, Crawford, and Alchian (1978). A very useful summary of these arguments is in Tirole's (1988) chapter on the theory of the firm. Explicit models of renegotiation include Grossman and Hart (1986), Hart and Moore (1988), and Dewatripont (1988).

The theory of incomplete contracts is discussed in Hart and Holmstrom (1987). Dye (1985) and Ayres and Gertner (1992) are examples of models of incompleteness based on explicit costs of contracting. Spier (1988) is a model of incompleteness based on strategic withholding of relevant information. The law and economics of relational contracts is analyzed in Schwartz (1992) and Hadfield (1990).

For an empirical study of price terms in long-term contracts, see Crocker and Masten (1991). Laffont and Tirole (1993) provides a clear technical discussion of the value of a commitment not to renegotiate. Sarig (1988) notes that dispersing the ability to renegotiate over diverse third parties is an effective way to make a commitment not to renegotiate.

CHAPTER 4

Signaling, Screening, and Nonverifiable Information

Signaling and Screening

The previous chapter focused on information that was verifiable or at least observable to the parties. In many instances, however, one party will possess private, nonverifiable information that neither the other party nor any third party can acquire directly. In such situations, legal rules cannot compel disclosure of the information. By definition, a court has no direct way of telling whether any disclosure is truthful. For this reason, a party obliged to disclose information has no reason to tell the truth. The best that uninformed parties or courts can do is draw inferences from the actions that the informed party takes. Legal rules, however, can have important effects nevertheless. By limiting the actions that parties can take or attaching consequences (such as an obligation to pay damages) to some actions, but not others, legal rules affect not only what actions are taken, but what inferences can be drawn from those actions.

The problems that arise when there is private, nonverifiable information have been a central concern of economists for the past several decades. Indeed, questions about the effects of private, nonverifiable information have been at the heart of economic analysis of such issues as dividend policy, capital structure, advertising, education, and insurance. The basic model we use in this chapter is therefore one that recurs in the analysis of these problems and many others.

The idea that some information is revealed only by the actions that individuals take has long been understood to have implications for lawmakers. King Solomon's judgment—testing someone's affection for a child by that person's unwillingness to allow the child to be injured—is the oldest and best-known story of dispute resolution. The problems from the perspective of a lawmaker are well known. The parent who is most aggressive in pursuing a child custody action may be the one who cares for the child the least. An ideal legal system must somehow take this into account.[1]

We can show how inferences can be drawn from actions with the following example. Those who apply for a job may know whether they are lazy or industrious, but those hiring them have no way of acquiring this information directly. The potential employers, however, may be able to draw inferences about the applicant's type from the actions that the applicant takes. Employers are most likely to be able to draw such inferences when there is an action that industrious applicants can take that is more attractive to them than to lazy applicants. What matters is not that the action be costless to the industrious workers, but that the lazy workers not take the same action. The action must be sufficiently unattractive to the lazy applicants so that they are unwilling to take it, even though the consequence of not taking the action is that they are identified as lazy workers. For the industrious workers, the benefits of distinguishing themselves from lazy workers have to be sufficient to offset the costs of taking the action.

Industrious workers, for example, may be able to *signal* that they are hard workers by completing a training program that lazy workers would find too taxing. Signaling takes place when those who possess nonverifiable information can convey that information in the way they choose their actions. Similarly, an employer may *screen* workers by offering a contract that is tied to output. Screening takes place when the uninformed players can choose actions that lead informed players to act in a way that reveals information. A lazy worker will find such a contract much less attractive than will an industrious worker. (Such a contract may have the further advantage of giving both kinds of workers an incentive to work harder.)

A legal rule requiring lazy workers to identify themselves cannot be enforced if a court has no way of telling whether someone is lazy. Legal rules can, however, affect the actions that individuals take, and thus affect the way information is transferred between parties. An insurance company may not be able to tell which potential customers are high-risk, but it may be able to screen different types of customers by offering different policies at different rates. A legal rule that regulates the terms of insurance policies affects an insurer's ability to screen potential customers and may thereby affect the ability of individuals to obtain insurance.

Legal rules can affect the ability of individuals to signal private information even when such rules are not directed at the problems that arise from nonverifiable information. Assume, for example, that buyers have no direct way of knowing whether a seller makes a high- or a low-quality product. High-quality sellers may be able to signal their type by selling goods with a warranty. Because their goods break down less often, these sellers can offer a warranty more cheaply than low-quality sellers. Low-quality sellers might be forced to sell without a warranty rather than try to sell with a warranty. Alternatively, the inability of low-quality sellers to offer goods with a warranty—and thus their inability to keep their identity hidden—may force them out of business altogether. The action of the high-quality seller (offering a warranty) conveys information to buyers.

When high-quality sellers can send this signal, low-quality sellers may decide to remain in the market but sell without a warranty. Buyers can choose between high-quality goods (with a warranty and at a higher price) and low-quality goods. Warranty coverage does not fully compensate the buyer for all the indirect costs that arise when goods fail. Hence, buyers still prefer high-quality goods to low-quality goods at a given price. In this environment, a legal rule mandating a warranty would have the effect of suppressing a signal. A legal rule requiring all goods to be sold with a warranty might prevent buyers from identifying high- and low-quality sellers and paying the appropriate price to each.

One should not infer from this example, however, that signals are necessarily good or that legal rules are bad if they suppress signals or limit the ability of others to respond to them. A warranty works for the high-quality seller only because that seller incurs fewer costs from sending the signal than a low-quality seller. Even if a signal is wasteful and inefficient, someone may send it because that person can send it more cheaply. Assume that firms that make high-quality products are more willing to advertise than firms that make low-quality products. Customers who buy high-quality products learn that they are high-quality after the purchase and are likely to buy the products again. By contrast, someone who buys a low-quality product because of advertising is unlikely to buy it again. Hence, spending money to persuade someone to buy a product initially brings a greater benefit to a high-quality seller than to a low-quality seller.

For this reason, buyers who observe that a seller is advertising can draw an inference that such a seller is more likely to be high-quality than a seller who is not advertising. Advertising is itself a signal of quality. We cannot, however, infer that advertising is good. The amount that high-quality sellers spend on advertising turns on the private benefit they enjoy from distinguishing themselves from low-quality sellers. This benefit may be much greater than the social benefit that is enjoyed when consumers can distinguish between high- and low-quality products.

Before we confront the merits of any particular rule affecting the ability of parties to signal information or to screen for it, we need to understand the different ways in which legal rules can affect how parties may be able to convey nonverifiable information to each other through their actions. In the next section, we show how these problems can be captured in an extensive form game.

Modeling Nonverifiable Information

The ultimate test of game theory as applied to the analysis of legal rules or anything else is whether it sheds light on how individuals are likely to behave. To take advantage of game theory, however, we do not need to recreate the exact process by which individuals make the decisions that they do. No one thinks that baseball players consciously solve quadratic equations whenever they throw the ball, but we can nevertheless predict the path the baseball takes using quadratic equations. The match between game theory and actual decisionmaking parallels this. Solution concepts such as the perfect Bayesian equilibrium assuredly do not model the process by which individuals make decisions. Nonetheless, the behavior of individuals may well be consistent with the behavior one would see if individuals made choices using the solution concepts of game theory.

To be sure, the more a model depends on the assumption that individuals behave in a way that is perfectly rational, the more suspect it is. After all, it is one thing to assume that individuals use heuristic devices that allow them to do the best they can with the resources they have and quite another to use models that make far stronger assumptions about rationality. One must be cautious about extending too far conclusions drawn from a model that depends crucially, for example, on all the players' being perfectly informed, not only about their own payoffs and their own beliefs, but also about those of others. We should, however, bear in mind that our ambition is not to produce exact prescriptions about the shape of legal rules, but to gain a basic understanding of how laws work. A rigorous methodology that generates models which map our world closely should give us insights into forces that are not readily apparent.

The easiest way to model nonverifiable information is to posit that the information has a binary character: workers are either lazy or industrious, and they each know which they are; sellers make high-quality or low-quality goods and know which kind; buyers of insurance know whether they are high- or low-risk. In these games, a hypothetical player, whom we can call *Nature,* has the first move. Nature decides whether the subsequent course of play will involve a lazy or an industrious worker, a high- or a low-quality seller, or a low- or a high-risk buyer. The effect of introducing Nature as a third player is to convert a game of incomplete information, which cannot easily be modeled, to one of imperfect information, which, as we saw in the last chapter, can readily be captured in extensive form. We can then solve these games using refinements of the perfect Bayesian equilibrium solution concept.

We shall look at games in which there are only two players in addition to Nature. Each player has one move and must choose between two actions. We then can set out the game in the extensive form. At every move, the uninformed player has an information set that contains two nodes. One node represents the decision that arises after Nature has chosen an informed player of one type (industrious, high-quality, low-risk), and the other reflects the decision that arises after Nature has chosen the other type of informed player (lazy, low-quality, high-risk). The private information has a similar binary character and concerns whether a player is of one type or another.

Like the two-by-two bimatrix in Chapter 1, this structure has the virtue of simplicity. It can capture the basic elements of strategic behavior in the face of private, nonverifiable information, but it is straightforward enough that one can readily see how legal rules transform these games and thus alter the behavior of the parties. In this section, we develop a game that has this structure and show how it is a useful way of understanding how legal rules can affect the way signals are sent and received when one player has private, nonverifiable information. As we have seen in earlier chapters, legal rules can change behavior dramatically, even when they affect only actions that do not appear in equilibrium.

Consider the game in Figure 4.1.[2] An employer must hire an additional worker. There are two types of workers: those with good backs and those with bad backs. The kind of back a worker has is private,

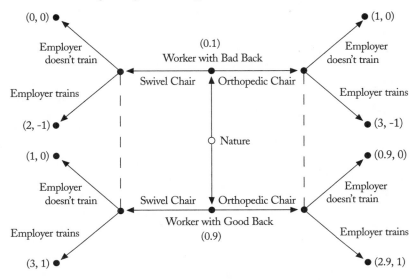

Figure 4.1 Signaling game (with private, nonverifiable information). *Payoffs:* Worker, Employer.

nonverifiable information. There is no way for an employer (or a court) to know directly whether a worker has a good or a bad back. The best an employer (or a court) can do is to draw inferences from the actions that a worker takes. Both the worker with a bad back and the worker with a good back must decide what kind of chair to request. It costs the employer the same amount to equip a worker with either kind of chair. Workers with bad backs put a value of $1 on having an orthopedic chair and a value of $0 on a swivel chair. Workers with good backs put a value of $1 on having a swivel chair and a value of $.90 on an orthopedic one.

After the worker picks a chair, the employer then decides whether to train the worker. In this game, we assume that the training is worth $2 to a worker. The training makes the job more enjoyable and provides the worker with skills that can be taken to another job. When the employer offers training to a worker with a good back, the employer benefits to the extent of $1 because the worker is more productive. Training a worker with a bad back, however, costs an employer $1. The employer gets fewer benefits from training workers with bad backs because they are more likely to be absent from work as a result of their back problem. The benefits to the employer of having a more productive worker do not offset the costs, given that the worker with a bad back is absent more often. As we have already noted, the employer is indifferent to which type of chair the worker requests, except to the extent that the request provides information about whether to offer training.

There are dashed lines between the two nodes after a worker has decided to ask for a swivel chair and between the two nodes after a worker decides to ask for an orthopedic chair. These show that the employer has no direct way of telling whether a given worker is one type or the other. Because the information is neither observable nor verifiable, workers cannot reveal their type. Workers with good backs would like to be identified as such. They as well as the employer are better off when the employer gives them training; hence, the workers with good backs want their employers to know their type. They stand in the same position as the sellers in the previous chapter who had learned that their houses were well built. Unlike the sellers, however, workers with good backs lack any means of revealing the information.

Although the information is not verifiable, the employer does have beliefs about the likelihood that a given worker has a good or a bad back. These beliefs are common knowledge: The employer has a certain set of prior beliefs about the different types of workers, the workers know these beliefs, the employer knows that they know them, and so forth. The employer's initial beliefs about the likelihood that a given worker has a bad back will probably be based on the actual percentage of people with bad backs in the relevant population. For this reason, the assumption of common knowledge is plausible. In this game, we assume that at the outset everyone shares the belief that there is a 90 percent chance that a worker has a good back and a 10 percent chance that the worker has a bad back.[3] (We indicate these beliefs in Figure 4.1 by the numbers we place next to each of the workers.) The players, however, update their beliefs in light of the actions that the different types of workers take in equilibrium.

In testing for a proposed equilibrium, we must ask whether the combination of actions and beliefs is consistent. As soon as we see that players would change their actions given the beliefs of another player, we know that we do not have a solution to the game, and we must look for another proposed equilibrium. That a deviation would occur in the wake of a particular belief in a proposed equilibrium shows us only that a particular combination of actions and beliefs cannot be an equilibrium. The desirability of deviating shows that an action of a player in the proposed equilibrium is not a best response, given the actions and beliefs of the others. The deviation itself does not tell us anything about what combination of actions and beliefs forms an equilibrium.

We can eliminate some strategy combinations immediately. For example, no equilibrium can have workers with good backs asking for orthopedic chairs and workers with bad backs asking for swivel chairs. To see this, we set out the employer's beliefs and actions in this proposed equilibrium and then ask if the workers are choosing a best response. The employer's beliefs also have to be consistent with the actions of the workers. Hence, in an equilibrium in which workers with good backs ask for orthopedic chairs and workers with bad backs ask for swivel chairs, the employer believes that everyone who asks for an orthopedic chair is a worker with a good back and that everyone who asks for a swivel chair is a worker with a bad back.

The employer makes the decision about whether to offer training accordingly. The employer does not offer training to workers who ask for swivel chairs. The employer's beliefs must be consistent with the actions of the players, and, in this proposed equilibrium, only workers with bad backs ask for swivel chairs. The employer must also take actions that provide the employer with the highest payoff given the belief. The employer will therefore not train these workers, thus receiving a payoff of $0 rather than −$1. Similarly, the employer will train workers who ask for orthopedic chairs (and, again under the proposed equilibrium, thereby reveal themselves to have good backs). In this way, the employer receives $1 rather than $0.

Once we set out these beliefs, we can ask if the action of each type of worker is a best response. In this proposed equilibrium, the action of the worker with a bad back is not a best response. Given the employer's belief that workers who ask for orthopedic chairs have good backs, the worker with a bad back would do better asking for an orthopedic chair. Because the employer believes that a worker who makes such a request has a good back, the worker with a bad back who asks for an orthopedic chair would receive training and enjoy a payoff of $3 rather than $0. This proposed combination of actions and beliefs in which workers with bad backs ask for swivel chairs therefore cannot be a solution to the game. The action of the worker with a bad back is not a best response given the beliefs of the employer, and in equilibrium it should be.

Using similar reasoning we can also reject as a solution to the game a proposed equilibrium in which workers with good backs ask for swivel chairs and workers with bad backs ask for orthopedic chairs. Again, in this combination of actions and beliefs, workers fully reveal their type by the kind of chair they request. The employer would therefore believe that anyone who asked for a swivel chair had a good back and anyone who asked for an orthopedic chair had a bad one. The employer would decide whether to offer training accordingly—and offer training only to workers who asked for swivel chairs. Once again, the action of the workers with bad backs would not be a best response. These workers would do better to ask for swivel chairs given that the employer believes that workers who ask for swivel chairs have good backs and therefore trains them. This proposed equilibrium also depends upon beliefs and actions that are inconsistent with each other.

There is no separating equilibrium in this game. This conclusion is entirely consistent with our intuitions about how the workers and the employer are likely to behave. The employer will never train a worker with a bad back who is identified as such. For their part, workers with bad backs value training at $2, but having the proper chair at only $1. Hence, they will never take any action involving the choice of chair that distinguishes them from workers with good backs. To do so would lead the employer to deny them training, and workers put a higher value on training than on having the proper chair.

Pooling equilibria—solutions in which both types of workers ask for the same kind of chair—do exist, however. One such equilibrium occurs when both types of workers ask for swivel chairs, the employer trains workers who ask for swivel chairs but does not train those who ask for orthopedic chairs, and the employer believes that (1) 10 percent of the workers who ask for swivel chairs have bad backs and the rest have good ones; and (2) anyone who asks for an orthopedic chair has a bad back. The action of the workers with good backs is a best response. These workers enjoy a payoff of $3, the highest they can receive in this game. The action of the workers with bad backs is also a best response, given the employer's belief that anyone who asks for an orthopedic chair has a bad back. Because employers believe that those who ask for orthopedic chairs have bad backs, they deny training to any worker who asks for an orthopedic chair. The worker with a bad back who asks for an orthopedic chair therefore enjoys a payoff of only $1. Such a worker is thus better off remaining with the equilibrium strategy of asking for a swivel chair and receiving a payoff of $2.

When both types of workers ask for swivel chairs, the employer's best response is to train everyone. All the workers take the same action. Hence, their actions contain no new information and the employer has no basis on which to revise the initial beliefs about the worker's type.

Bayes's rule requires that the employer's belief remain unchanged after observing an action that conveys no new information. Therefore, the employer believes that the workers who ask for swivel chairs have the same proportion of good and bad backs as the population at large. When the employer trains all the workers, the employer receives a payoff of $0.80. The employer would receive a payoff of $0 if the employer did not train them.[4]

In this proposed equilibrium no one asks for orthopedic chairs. The actions of the players do not impose any restrictions on the beliefs that the employer has about someone who requests an orthopedic chair. The existence of this pooling equilibrium, however, depends on the employer's belief that at least ½ of those who ask for orthopedic chairs have bad backs. If the employer thought that those who asked for orthopedic chairs were workers with good backs, the employer would train anyone who made such a request. For those with bad backs to ask for swivel chairs in this proposed equilibrium is not a best response. They would do better asking for orthopedic chairs, given that the employer would still train them.

We have not offered any reason why, in an equilibrium in which everyone asks for swivel chairs, the employer should believe that those who ask for orthopedic chairs are likely to have bad backs. This belief (that those who ask for orthopedic chairs have bad backs) happens to be one that supports the pooling equilibrium in which everyone asks for a swivel chair and everyone receives training. Nothing about the perfect Bayesian solution concept, however, suggests that an employer is likely to have this belief or that the employer should have it. Because no one asks for these chairs in equilibrium, Bayes's rule does not constrain the employer's beliefs. Any belief is possible.

So far we have found one pooling equilibrium. Just as games may have multiple Nash equilibria, a game may have multiple perfect Bayesian equilibria. This game does. Consider the following combination of actions and beliefs. Both types of workers ask for orthopedic chairs and the employer trains workers who ask for orthopedic chairs but does not train those who ask for swivel chairs. The employer believes that 90 percent of those who ask for orthopedic chairs have good backs, and the employer also believes that, if anyone asked for a swivel chair, that person would probably be a worker with a bad back. In this proposed equilibrium, workers with bad backs would enjoy a payoff of $3, the maximum they can receive in this game. Hence, they have no incentive to deviate. Workers with good backs enjoy a payoff of $2.90. They would enjoy a payoff of $3 if they asked for a swivel chair and the employer gave them training, but the employer's beliefs are such that the employer will not train anyone who asks for a swivel chair. (The employer believes that those who ask for swivel chairs are workers with bad backs and hence would not train anyone who asked for one.)

The employer's belief that 90 percent of those who ask for orthopedic chairs are workers with good backs is consistent with Bayes's rule. In this proposed equilibrium, no action is taken that provides the employer with any new information about the worker's type. The employer's initial belief should thus remain unchanged. Given that the employer's initial belief was that 90 percent of all workers have good backs, and given that all workers in this equilibrium ask for orthopedic chairs, the employer is better off training employees than not. The employer's belief that most of those who ask for swivel chairs have bad backs is off the equilibrium path. Nothing in the perfect Bayesian equilibrium solution concept requires the employer to have this belief, but nothing rules it out either. Our intuitions tell us that it is implausible that the employer should have this belief, but before we can look at how legal rules work in this environment, we need to find a formal way to explain why such a belief is implausible.

The employer's belief that those who ask for swivel chairs have bad backs survives the refinement of the perfect Bayesian equilibrium that we introduced in the last chapter. (Recall that we posited that a combination of strategies and beliefs is implausible if it depends upon the belief that the other player adopts a dominated strategy.) Even if we constrain the employer's beliefs in this way, the employer can still believe that those who ask for swivel chairs are likely to have bad backs in an equilibrium in which all workers ask for orthopedic chairs. The maximum possible payoff that a worker with a bad back can receive by asking for a swivel chair—a payoff of $2—exceeds the minimum payoff that such a worker can receive by asking for an orthopedic chair (a payoff of $1). Hence, the strategy of asking for an orthopedic chair does not dominate the strategy of asking for a swivel chair for the worker with a bad back.

There is, however, a basis on which we can challenge the employer's belief about behavior off the equilibrium path. At the outset, the employer places only a 10 percent probability on any worker's having a bad back. If the employer were to see a worker asking for a swivel chair, how should the employer update that assessment? A worker with a bad back receives a $3 payoff in the proposed equilibrium. This $3 payoff dominates any payoff that the worker with a bad back could hope to receive by deviating ($0 or $2). This observation suggests a way in which one can refine the perfect Bayesian equilibrium solution: *An uninformed player's beliefs about behavior off the equilibrium path should be such that this player believes that, if a deviation were to take place, the player who deviates would, if at all possible, not be of a type whose payoff in the proposed equilibrium is larger than any payoff that player could ever receive by deviating.* This refinement of the perfect Bayesian equilibrium is known as *equilibrium dominance.* Just as a player should not expect another to play a dominated strategy, a player should not expect a player to choose an action that yields a payoff lower than the one that the player receives in the proposed equilibrium. This refinement is also called the *intuitive criterion.*[5]

We can test the reasonableness of the employer's beliefs in this proposed equilibrium another way. A worker with a good back wants to deviate but cannot, given the employer's belief that those who ask for swivel chairs have bad backs. Such a worker should be able to persuade the employer to change the belief. The worker with a good back could say, "If you were to see someone deviating, why should it be someone with a bad back? A worker with a bad back would be giving up a payoff of $3 in the proposed equilibrium in order to receive a payoff of $2 or $0. You should rather believe that the person who is deviating has a good back. Such a worker receives a payoff of $2.90 in the proposed equilibrium. A worker with a good back has at least some chance of doing better by deviating, because there is a chance that such a worker could receive $3." The employer would probably find such an argument persuasive. We do not, however, need to posit that the players can actually communicate with each other. If we can imagine such a conversation, the employer should be able to imagine it as well. The employer should not hold beliefs in equilibrium that would change in response to reasoned argument. Hence, we should reject this proposed equilibrium in which both types of workers ask for orthopedic chairs as a solution to the game.

We return to the equilibrium in which both workers ask for swivel chairs. Recall that this equilibrium depends upon the employer's belief that a worker who asks for an orthopedic chair has a bad back. We can now explain why the employer is likely to have this belief. No matter how the employer responds when a worker asks for an orthopedic chair, workers with good backs do worse by asking for an orthopedic chair than they would if they stuck with the equilibrium strategy and asked for a swivel chair. The employer should therefore infer that anyone who asks for an orthopedic chair has a bad back. Because the employer has this belief, the best response of workers with bad backs in this proposed equilibrium is to mimic the workers with good backs and ask for swivel chairs. The solution to this game is the pooling equilibrium in which everyone asks for a swivel chair and the employer trains everyone. The employer believes both that 90 percent of those who ask for swivel chairs have good backs and that any worker who asks for an orthopedic chair has a bad back.

This equilibrium is the only one in the game that survives the equilibrium dominance refinement of the perfect Bayesian equilibrium solution concept. Workers with bad backs value training more than having a suitable chair. Moreover, they recognize that employers will not train those they know to have bad backs, but will train everyone if they cannot determine who has a bad back. The refinement of the perfect Bayesian equilibrium concept that we used to solve this game does depend on a number of strong assumptions, such as common knowledge. Nevertheless, this refinement generally does a good job of isolating how parties are likely to behave when they must draw inferences from the actions of others. Moreover, it provides a formal way to think about how individuals may hide information by mimicking the actions of others, even though everyone as a group might be better off if they did not. We can use it to gain an understanding of how legal rules are likely to affect the way individuals behave when one possesses private, nonverifiable information.

The equilibrium in which everyone asks for a swivel chair brings a total benefit of $3.70.[6] This outcome falls short of the socially optimal outcome. (When all workers receive training and each type receives the appropriate chair, the total payoff is $3.80.)[7] In the socially optimal outcome, the workers with bad backs are better off, but neither the employer nor the workers with good backs are worse off. (The employer trains the workers with bad backs in both cases, so its position is unchanged, and the workers with bad backs are made better off by swapping the swivel chair for the orthopedic chair.) The workers with bad backs, however, do not want their employer to be able to draw any inferences about the type of back they have. If the employer could, the employer would refuse to train them.

Even if the employer and the worker could bargain with each other, this inefficient equilibrium would still exist. The employer cannot bribe the workers with bad backs and thereby induce them to reveal the information. The training brings a $2 benefit to the workers. This benefit exceeds the $1 it costs the employer to provide it. The employer is better off training the workers than trying to offer them enough money so that they reveal their type and forgo training. Hence, the employer has no incentive to alter the equilibrium.

Workers with bad backs have nothing to gain from revealing the information either. These workers would have to pay the employer $1 each to receive training after revealing the information, an amount equal to the benefit they receive from having the orthopedic chair. Hence, they too have nothing to gain from bargaining with the employer and thereby altering the equilibrium.

The outcome in which all workers have swivel chairs and are trained is not even second-best. When all workers are trained and receive orthopedic chairs, total benefits are slightly larger ($3.71).[8] In this case, however, those with good backs are worse off, whereas those with bad backs are better off. When all workers have swivel chairs, workers with good backs receive $3, workers with bad backs receive $2, and the employer receives an expected profit of $0.80. When all workers have orthopedic chairs, workers with good backs receive $2.90, workers with bad backs receive $3, and again the employer receives an expected profit of $0.80. Moving from swivel chairs to orthopedic ones brings a loss of $0.10 to workers with good backs and a gain of $1 to workers with bad backs. Workers with good backs form 90 percent of the population, and their expected loss from the shift is $0.09. Workers with bad backs form only 10 percent of the population, making their expected gain from the shift $0.10. The large gains to each of the individual workers with bad backs are slightly greater than the small loss to each of those with good backs. The employer is indifferent between either outcome and enjoys an expected benefit of $0.80 in both cases.

Even if courts cannot identify which workers have bad backs, workers as a group in this game would be made better off if a legal rule required employers to furnish orthopedic chairs to all workers.[9] The effect of such a rule is to change the equilibrium outcome of the game from one in which all workers choose swivel chairs to one in which they all choose orthopedic ones. Once the employer is required to supply orthopedic chairs to everyone, the workers with bad backs can receive training without revealing any private information. Hence, they would still be trained.

Under the assumptions in this model, a legal rule that prevented one type of worker from sending a signal about private information changed the outcome of the game from one pooling equilibrium to another. In games with nonverifiable information, such restrictions can enhance social welfare. In the absence of such a legal rule, the minority of workers who have bad backs cannot afford to be identified as such and thus must make themselves resemble the other workers. Workers with good backs will ask for swivel chairs and enjoy a payoff of $3 rather than $2.90. This action on the part of workers with good backs forces the workers with bad backs to ask for swivel chairs as well, even though they would enjoy a $1 gain from orthopedic chairs. They take this action because they would rather suffer with the swivel chair than identify themselves as workers with bad backs and thus deprive themselves of the benefits of training. Without legal intervention, the act of asking for an orthopedic chair would signal that the worker asking for it had a bad back. As a result, a pooling equilibrium emerges in which everyone asks for a swivel chair. There is no assurance that such an equilibrium is more desirable than one in which, by virtue of a legal rule, everyone used an orthopedic chair.

A legal rule that changes the solution of the game from one pooling equilibrium to another falls short of the first-best outcome of this game. In both cases, some workers do not have the chairs they value the most. Nevertheless, having a legal rule limiting the signals that workers can send leads to a better outcome than not limiting the actions of players at all. This example illustrates one way in which a legal rule can enter an environment in which there is private information and, in theory, change things for the better. Under slightly different numbers, such a legal rule could change things for the worse. If, in our example, the loss that workers with good backs suffered in moving from swivel chairs to orthopedic ones was $0.15 instead of $0.10, the pooling equilibrium in which all workers had swivel chairs would bring greater benefits than one in which all workers had orthopedic chairs. Mandating orthopedic chairs under these conditions would move us from a more efficient equilibrium to a less efficient one.

In addition to imposing restrictions on the actions of the party who possesses private information, legal rules can limit the actions of the person who is uninformed. If, as we have been assuming, the employer's beliefs at the start of the game are that most workers have good backs, the only equilibrium that survives refinements is one in which all workers ask for swivel chairs. As long as the employer will train all workers when those with bad backs cannot be identified, any pooling equilibrium is inefficient relative to the separating equilibrium that would exist if the law required training.

In the context of this model, a law that forbids discrimination in the provision of training makes the workers with bad backs better off and leaves the employer no worse off. A legal rule requiring training or banning discrimination allows workers with bad backs to ask for orthopedic chairs and still receive training. The workers can ask for the chair they want when the employer lacks the ability to act on the inferences that can be drawn from their actions. Stripping away the employer's unexercised right not to train workers does not leave the employer any worse off. Because workers with bad backs hide their type, the employer trains them even when discrimination is permitted. The employer's inability to distinguish between workers makes the right to discriminate irrelevant. A law mandating training requires the employer to do something that the employer would do anyway.

We can extend this observation to the more general case in which we do not know the relative value of the benefits of training to workers with bad backs or the cost of that training to the employer. When all workers ask for swivel chairs in equilibrium and the employer trains all of them, a legal rule requiring the employer to train workers (or not to discriminate in the provision of training) is desirable. It does not matter if the cost to the employer of training workers with bad backs exceeds the benefit to the workers with bad backs of having the training. As long as the employer trains workers with bad backs in the pooling equilibrium that arises in the absence of a ban on discrimination, we are better off imposing a ban that allows the two types of workers to separate. The employer is no worse off, and the workers with bad backs no longer incur the costs associated with trying to mimic workers with good backs.

At this point, we can compare rules that limit the ability of informed players to take actions that convey information (such as asking for a particular kind of chair) and rules that limit the ability of uninformed players to respond (for example, by deciding whether to offer training). If we face a pooling equilibrium in the existing legal regime, we may make the parties better off by limiting the actions of the informed players and thereby move from one pooling equilibrium to another. Whether the change is an improvement is determined by the differences between the two equilibria. In our example, the question was whether the total benefit to both types of workers of having orthopedic chairs exceeded the total benefit to both types of workers of having swivel chairs. (We do not need to take account of the employer, who receives the same payoff in both equilibria.)

Whenever we move from one pooling equilibrium to another, however, we face a trade-off. One type of player is better off and another is worse off. Restrictions on an uninformed player's ability to respond to signals are different in an important respect, for they may not require such a trade-off. If the uninformed player would not play a strategy anyway, banning the choice of that strategy would impose no extra cost on the uninformed player and would allow the informed player to choose without considering how the choice could convey information to the other player about the informed player's type.

In earlier chapters our focus was primarily on rules of civil damages that kept the total payoffs under any combination of strategies constant but shifted the share that each player received. So far in this chapter, however, we have examined legal rules that limit the actions that the players might take. These legal rules affect the strategy space of the players but not the payoffs. We should now return to civil damages and the question of how these work when information is nonverifiable. If courts were able to determine whether a worker had a bad back, we could generate the optimal outcome by requiring the employer to pay workers with bad backs the benefit they would enjoy if they received training in the event that the employer chose not to train them. We can also create a damage rule that does not require a court to identify workers with bad backs.

A rule requiring the employer to pay damages would force the employer to internalize the cost of the training decision. Consider a legal rule in which the employer had to pay a worker who asked for an orthopedic chair the value of the training to that worker if the employer did not actually provide the training. This rule would transform the game in Figure 4.1 to that in Figure 4.2. This game has two perfect Bayesian equilibria: One is the pooling equilibrium in which both workers ask for orthopedic chairs and receive training; and the other is the separating equilibrium in which the workers with bad backs ask for orthopedic chairs, workers with good backs ask for swivel chairs, and both receive training.

We can reject the pooling equilibrium in which both workers ask for orthopedic chairs. It requires the employer to believe that those who would ask for swivel chairs would be workers with bad backs. These beliefs about actions off the equilibrium path are implausible because the employer should not believe that a worker will choose a dominated strategy. (Asking for an orthopedic chair gives workers with a bad back a payoff of $3 whether they receive training or not, whereas asking for a swivel chair can never bring more than $2.) The second equilibrium survives refinements and therefore is the only reasonable solution to this game. This civil damages rule—one that is geared to a visible action and thus one that a court can enforce—induces separation and therefore maximizes overall social welfare.

This rule also induces workers with bad backs to ask for the chair

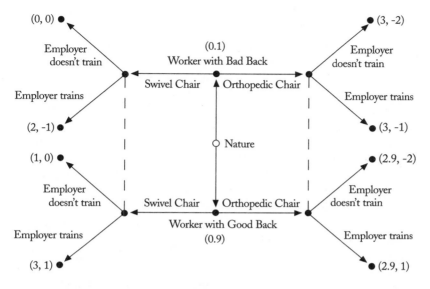

Figure 4.2 Signaling game (with nonverifiable information and civil damages). *Payoffs:* Worker, Employer.

that suits them, thus revealing their type to the employer. In the absence of such a rule, workers with bad backs focus only on the private benefits accruing from their decision when they choose whether to mimic workers with good backs by asking for swivel chairs. They compare only the benefits of training and the benefits of having an orthopedic chair. In the absence of a civil damages rule, workers with bad backs will not take account of the potential benefits the employer receives from knowing whether a worker has a good or a bad back. With this rule in place, the worker with a bad back has nothing to gain from asking for a swivel chair because the legal rule ensures that the worker will receive either training or its dollar equivalent. The employer in turn decides to train on the basis of whether the training brings benefits to the employer and the worker jointly that exceed the cost of the training.

Implementing a civil damages rule in the context of this model requires only that we know the value of the training to the worker with the bad back. Civil damages is a form of decentralized rule, which may be preferable to a centralized rule that could require us to know more at the outset. (We cannot, for example, be confident that we are improving the outcome by mandating that all workers be given orthopedic

chairs unless we know the value of each chair to both types of workers.)
Even if compensation rules are informationally more parsimonious, however, they require repeated legal supervision, whereas the
command-and-control rules require just an initial decision.[10]

In the absence of a legal rule limiting the actions of the players or
imposing damages on the employer, the actions of workers and the
employer generate an equilibrium in which all workers ask for swivel
chairs. The payoffs in the game are such that training is more valuable
than having one kind of chair or the other. Because workers with good
backs are in the majority, they are able both to choose the chair that
brings them the greatest benefit and to enjoy training. The workers
with bad backs are in the minority and must mimic workers with good
backs in order to avoid identification and receive training. We next
turn to a game in which separation occurs without legal intervention
and in which the welfare effects of intervention once again depend on
the relevant parameter values. In this game, all the payoffs are the same
as in Figure 4.1, except that the value to the workers with bad backs
of having orthopedic chairs is $3 instead of $1. This game is illustrated
in Figure 4.3.

This game has two perfect Bayesian equilibria. The first is a separat-

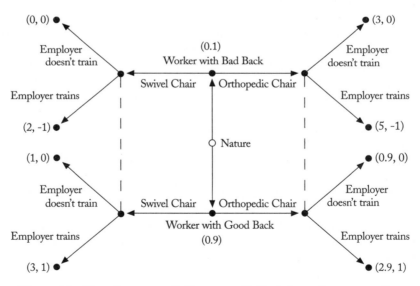

Figure 4.3 Signaling game (with nonverifiable information; separating
equilibrium). *Payoffs:* Worker, Employer.

ing equilibrium. The workers with good backs ask for swivel chairs and the employer trains them, whereas the workers with bad backs ask for orthopedic chairs and the employer does not train them. Confirming that this is a perfect Bayesian equilibrium is straightforward. If only workers with good backs ask for swivel chairs in equilibrium, then the employer believes that those who ask for swivel chairs have good backs and those who ask for orthopedic chairs have bad ones. We now ask if the action of the workers is a best response, given this belief. Workers with bad backs receive $3 when they ask for orthopedic chairs. This amount is greater than the $2 they receive after asking for swivel chairs. (They receive $2 because the employer in this proposed equilibrium believes that those who ask for swivel chairs have good backs.) In this proposed equilibrium, for workers with good backs to ask for swivel chairs is a best response—they receive $3 in this equilibrium but would receive only $.90 if they asked for orthopedic chairs.

The second perfect Bayesian equilibrium in this game is a pooling equilibrium in which both types of workers ask for orthopedic chairs. This pooling equilibrium, however, requires the employer to believe that any worker who asks for a swivel chair has a bad back. This belief is not plausible. Workers with bad backs are always better off asking for an orthopedic chair, no matter how the employer responds. The employer can believe that those who ask for swivel chairs have bad backs only by believing that workers with bad backs would play a dominated strategy. Workers with good backs therefore receive a payoff of $3 by asking for swivel chairs and a payoff of only $2.90 by asking for orthopedic chairs. Because asking for a swivel chair is not a dominated strategy for them, the employer should believe that those who ask for swivel chairs have good backs. Once the employer has this belief, there cannot be an equilibrium in which both types of workers ask for orthopedic chairs.

Under these parameter values, total social welfare in the separating equilibrium is $3.90, but in the pooling equilibrium in which everyone asks for orthopedic chairs it is $3.91. Again, if requiring orthopedic chairs were costless, legal intervention would push us to a better equilibrium. But tweaking the parameters slightly gives us a different result. Requiring orthopedic chairs could easily reduce social welfare. There is a difference between this game and the one shown in Figure 4.1 that is worth noting. In the game in Figure 4.1, a ban on the actions of the informed player who sent the signal—in that case, a ban on asking for swivel chairs—shifted us from one pooling equilibrium to another. In Figure 4.2, however, such a ban converts a separating equilibrium into a pooling equilibrium. By definition, less information is communicated in the pooling equilibrium. The employer is unable to identify different types of workers and cannot make training decisions in light of the differences. As before, workers ignore the external effects of their decisions on the employer.

These variations on a single signaling game involving an employer and workers with good and bad backs show the ways in which legal rules can alter the ability of individuals to convey private, nonverifiable information to one another through their actions. We apply the basic lessons of this model in a number of different contexts in the next section. Before we do this, however, a few basic points are worth repeating. When lawmakers confront a problem of private, nonverifiable information, they face a fundamentally different problem from the one we saw in the last chapter. Because the information is nonverifiable, unraveling will not naturally occur. A court confronting nonverifiable information is necessarily left in the dark as well. The best a lawmaker can do is to alter the behavior of the parties by attaching consequences to actions that a court can verify.

In some cases, a first-best outcome is not possible and the choice is between one pooling equilibrium and another. In other cases, the first-best outcome may be possible, but in these cases the appropriate rule may be one that forbids the uninformed player from taking actions that would never take place in equilibrium anyway. Such a rule allows the informed player to take an action that previously would not have been taken because the action would have revealed the informed player's type, with adverse consequences. Understanding the kinds of equilibria that are plausible and what supports them becomes crucial when legal rules work by limiting actions that are off the equilibrium path.

Signals and the Effects of Legal Rules

The model we developed in the last section provides a framework for analyzing many other kinds of problems that legal rules address. Consider, for example, a law that Congress enacted in 1993 governing parental leave.[11] No one doubts that a parental leave is a good idea when both employer and employee agree to it. The question is whether the law should require them in situations in which the parties would not agree to them otherwise. One can argue that employers would offer leaves if workers valued them more, say, than a correspondingly higher salary. Mandating leaves when the parties do not agree to them voluntarily might make the parties worse off. Indeed, if some employers find offering leave cheaper than others, we may have the best possible outcome by allowing employees who want parental leave to seek out employers who are best able to offer it. One can respond to this line of argument a number of ways. We want to ask whether contractual agreements might fall short of what is in the interests of both parties because of the existence of private, nonverifiable information.

Potential employees may not want to bargain for parental leave because of the other inferences (legitimate or illegitimate) that the employer might draw about them before hiring them. Rather than asking for parental leave, these potential employees may mimic the behavior of those who do not want leaves to avoid signaling that they were going to become parents before they were already hired. A pooling equilibrium emerges in which no one asks for a leave. As we saw in the last section, such pooling equilibria are necessarily inefficient and may be less desirable than other pooling equilibria that the law can create. The law that requires employers to offer parental leave could be an example of a law that transforms a less efficient equilibrium into a more efficient one. We can get a sense of how this might be the case by returning to the basic structure of the model we developed in the last section.

Let us assume that there are two types of prospective workers: those with a low probability of becoming parents and those with a high probability. There is only one kind of employer.[12] To cast the problem in its starkest form, assume further that this information is private and not verifiable. Those with a high probability of becoming parents are willing to have a parental leave clause built into their contracts in exchange for a lower wage. Absent information problems, the employer would agree to such a deal. The costs to the employer of offering such terms are lower than the benefits to the prospective parent. A contract with parental leave can make both parties better off. By contrast, employees with a low probability of becoming parents prefer a higher wage and no parental leave contract.

The problem changes, however, if we make the further assumption that the employer would rather hire workers who will not become parents for reasons unrelated to parental leave. The employer might think, for example, that workers who become parents will not work as hard or will be less willing to work overtime after they become parents. As long as the employer has this belief, potential workers must take it into account when they negotiate their employment contracts. For this reason, workers who want parental leave might mimic the strategy of the other type of worker, thereby increasing their chances of being hired in the first place or increasing their wage. In the last section, workers with bad backs in the model mimicked workers with good backs by asking for a chair that they did not prefer. In this case, workers who are likely to become parents might enter into a suboptimal contract. These workers may be worse off if they disclose this information, given the attitude of their employer—not toward parental leave per se, but rather toward those who become parents.

If asking for parental leave disclosed no information, workers who were likely to become parents could ask for a contract that included such a clause. The signal that asking for such a clause sends, however, may keep prospective parents from raising the issue. This outcome— in which all workers pool and have contracts without parental leave clauses—may be the worst of the two pooling equilibria. In the previous section, workers with bad backs strongly preferred orthopedic chairs, whereas workers with good backs slightly preferred swivel chairs. Similarly, the benefit to prospective parents of having such a clause may be large and the cost (presumably a lower wage) to workers who are unlikely to become parents may be small.

If we require parental leave as a term of the contract, we can shift from a pooling equilibrium in which no contract offers parental leave to one in which all do. Such a mandatory rule may be an improvement over a regime in which parental leave is left entirely to private contracting. Nevertheless, this rule may fall short of the ideal. In this pooling equilibrium, the contract is not tailored to the needs of the workers who are not likely to become parents. The first-best solution may be one in which workers ask for parental leave or not, depending on their type, and in which the employer does not take the probability of becoming a parent into account in making the hiring decision or any other terms of employment.

A law that could, at low cost, prevent the employer from taking the information into account might be an improvement over either no law limiting what the employer could do or over one requiring a mandatory term. Such a law could have the same effect as the law that prevented employers from discriminating against workers in deciding whether to train them. Employers cannot discriminate against those who are likely to become parents if those individuals never reveal themselves. Hence, a law that bars such discrimination may not leave employers any worse off. Those workers who are not likely to become parents are in the same position as they were when no contracts offered parental leave, and those who are likely to become parents are now better off because they can bargain for parental leave. They no longer need to fear that they will suffer other consequences by revealing that they will probably become parents. Once assured that the employer will not discriminate in hiring, both types of workers are willing to reveal their type and bargain for the contract that accommodates their needs.

The possible existence of private, nonverifiable information provides no justification for statutes banning discrimination on the basis of race or sex. In these cases, the employer can readily observe the relevant characteristics. Workers cannot hide their type, and so their actions will not reflect a desire to keep it hidden. The justification for these laws lies elsewhere. The scope of antidiscrimination laws, however, extends beyond characteristics that are readily discernible or verifiable. Some physical and mental disabilities that fall within the scope of antidiscrimination laws are not verifiable information from the perspective of a potential employer. Similarly, sexual orientation may not be verifiable to an employer either. In both cases, the employer might, in the absence of antidiscrimination laws, draw inferences from actions that workers take. Without a legal regime in place, workers who are the potential victims of discrimination might have to incur significant costs in order to mimic the actions of the dominant group.

Introducing any legal rule introduces costs, but this analysis suggests that the costs of rules aimed at discrimination may differ depending on their type. A rule that mandates particular actions or particular contract terms imposes a cost on the party who would, in fact, bargain for a different term. Limits on the actions of uninformed players may, by their nature, be more difficult to enforce. It is easier to tell whether an employer is providing an orthopedic chair to everyone than it is to tell whether an employer is offering the same training that would be offered if the employer did not know that a worker had a bad back. In addition, such rules can have other collateral consequences when enforcement is costly. Rather than risk facing a claim that discrimination occurred in the kind of training that was offered, an employer might not offer any at all.

Laws that impose mandatory terms can also destroy an efficient separating equilibrium. Such laws can inhibit the flow of privately held information. One example may be plant closing laws; that is, laws that require employers to give workers advance notice of any closing.[13] We can again illustrate the point by creating a model whose basic structure tracks that of the game involving the employer and the two types of workers. A firm operates a plant. The plant may succeed or fail. If the plant fails and is closed, its workers will have to find new jobs. Workers incur costs from losing their jobs if the plant closes. The world is populated by two types of firms. One type, the low-risk firm, is unlikely to close its plant. The other type, the high-risk firm, is much more likely to do so. Workers care about the type of firm for which they work. Other things being equal, workers would prefer to work for the low-risk firm, as the expected job loss costs associated with it are lower than they are with the high-risk firm.

The costs associated with job loss can be influenced by both the firm and the workers. If the firm gives advance notice of closure, it will reduce the job loss costs. Job loss costs are also reduced if the workers monitor the firm and detect that it will close. Monitoring and disclosure directly affect worker productivity and therefore firm profitability. If the workers monitor the firm, productivity is reduced. If the firm announces that it will close, productivity is also reduced, in addition to any productivity losses from the workers' monitoring. If the firm does not disclose the information but the workers detect closure, productivity is also reduced. The firm can influence whether the worker will want to monitor by voluntarily agreeing to give advance notice of a closing. The ability to offer such a term in the employment contract may itself be an efficient signal that allows low-risk firms to distinguish themselves from high-risk firms. When firms have the ability to offer such terms, workers may be better off. They no longer have to monitor low-risk firms because they will be given notice; and they can demand a higher wage from the high-risk firms who reveal their type by refusing to offer such a term in their employment contract.

Offering a contract with or without a notice provision is analogous to asking for a swivel or an orthopedic chair. The worker's decision to monitor is analogous to the employer's decision about training. The payoffs may be structured in such a way, however, that, in the absence of any legal rule, there is an efficient separating equilibrium in which only low-risk firms have notice provisions and workers monitor only high-risk firms. Workers have no direct way of knowing that a firm is high-risk, but they can infer it from the failure of the firm to offer a notice provision in its contract.

A law requiring disclosure of plans to close might destroy this separating equilibrium. In the absence of such a law, low-risk firms might distinguish themselves from high-risk firms by voluntarily accepting the duty to disclose in their labor agreements with their workers. When a law requires both low-risk and high-risk firms to give advance notice, however, the low-risk firm might no longer be able to distinguish itself. The separating equilibrium that might otherwise result will be destroyed and will instead be replaced by a pooling equilibrium in which no information about the type of the firm is communicated. Firms and workers as a group will be worse off. Without the information, workers will attempt to minimize their anticipated job loss costs based upon the average type rather than upon the actual type, something they would know in a separating equilibrium.

More generally, the willingness of a party to agree voluntarily to a term in a contract may signal the party's type. Imposing a mandatory term may prevent this signaling and thereby reduce the amount of information transferred. To put the idea differently, contracts must be written on observable events. Disclosing or not disclosing that the firm is going to shut down is such an event. A contract term based on an observable event can communicate information only if parties have a choice about whether to include such a term in their contract. Every mandatory term potentially brings with it a hidden cost because it may prevent parties from revealing nonverifiable information to one another.

Information Revelation and Contract Default Rules

The vast majority of contract rules are not mandatory requirements like the ones discussed in the preceding section.[14] Rather, they are *default rules*, rules that come into play only if the contract is silent with respect to the relevant contingency. Contracting parties are, for example, able to limit the damages that can be recovered from breach of contract if they choose. In this section, we discuss how default rules can affect information revelation in ways similar to mandatory clauses.

We begin with the classic case of *Hadley v. Baxendale*.[15] In this case, a miller contracted with a carrier to transport a broken crankshaft from Gloucester to Greenwich. The shipment was delayed. The miller sued and the court had to decide whether the carrier should be liable for the profits lost during the increased time the mill was shut down. The court ruled that, because the miller had not told the carrier about its special needs, it could not recover the profits lost from the delay. These damages were not reasonably foreseeable.

Modern contract law has generally adopted the rule that a person is not entitled to consequential damages that are not reasonably foreseeable. As applied to modern shipping contracts, the default rule that emerged in *Hadley* may be neither sensible nor very important.[16] Our interest, however, is not in the details of this case itself, but in how default rules work in situations in which problems of information and strategic behavior loom large.

Let us assume that there are two types of millers. One type of miller is low-damage. In the event that the carrier fails to deliver the crank-shaft on time, the low-damage miller suffers damages of $100. The other type of miller is high-damage and suffers $150 in damages when the carrier fails to deliver the crankshaft on time. The carrier knows that 20 percent of the millers are high-damage and that 80 percent are low-damage, but the carrier has no way of telling one type from the other. Nor do low-damage millers have any direct means of revealing their type. If a miller did its own shipping, the amount of care the miller would take would depend on its type. By spending more money on transporting the goods, one can reduce the likelihood of an accident. Hence, if the miller were transporting the goods itself, a high-damage miller would take more care and a low-damage miller less. We need one more assumption before we proceed: Prices reflect the cost of shipping and the expected costs of damages.

As a first step to the analysis of default damage rules, let us consider the effects of different damage rules in cases in which the parties cannot contract around the rule. A consequential damages rule makes the carrier liable for $150 if it does not perform and the miller is high-damage, and $100 if it does not perform and the miller is low-damage. If there are no actions that the low-damage millers can take to signal their type, there will be a pooling equilibrium. The carrier will take precautions consistent with the average of the two possibilities.

By contrast, a foreseeable damages rule would oblige the carrier to pay all millers only $100 in the event of breach, regardless of whether they were high- or low-damage millers.[17] Neither of these outcomes is desirable because the level of precaution does not correspond to the level of damages that a specific miller will suffer. A consequential damages rule may well be superior to a foreseeable damages rule because choosing precaution consistent with the average level of damages is likely to be better than using the correct level of precaution for low-damage millers, but underinvesting significantly for high-damage millers.

This game differs from those we looked at earlier in the chapter. The previous cases were ones in which private, nonverifiable information had ancillary effects. Unlike workers who care about both getting the correct chair and receiving training, the players here care only about the shipping of the goods. Workers with bad backs want to keep their employer from learning about their type after receiving orthopedic chairs. This information affects whether the employer trains them. By contrast, the high-damage millers do not care if the carrier learns their type after the contract for shipping the mill shaft has been signed. In the absence of such ancillary effects, parties always find it in their interest to negotiate around a default rule if transaction costs are low enough. The default rules matter only because there may be significant transaction costs.

Under a foreseeable damages rule, the high-damage miller does quite poorly because the carrier takes the precaution that is optimal for a low-damage miller. It may pay for the high-damage miller to negotiate around the default rule, ask for higher damages, and pay the carrier for this increased "insurance" against nonperformance. The benefits to the high-damage miller from such insurance outweigh the costs to the carrier of providing it because of the extra care that the carrier can take. There are no other ancillary effects from being a high-damage miller that the high-damage miller wants to conceal. Hence, the high-damage miller has an incentive to negotiate around the default rule. When the high-damage miller does this, both millers enjoy the contracts that are best for them.

The consequential damages rule, in contrast, leads to the optimal outcome only if low-damage millers negotiate around the default rule. Both types of millers are fully insured against nonperformance, but the price paid for the insurance is wrong. The price of the insurance is based on the expected damages that the carrier will pay. Because the high-damage miller is paid more than the low-damage miller under a consequential damages rule, the low-damage miller pays too much for shipping and the high-damage miller too little. The benefits from bargaining around the default rule are therefore concentrated on low-damage millers, who would want to negotiate around the default rule as well as agree to liquidated damages of $100 and a lower price for shipping. When the low-damage millers signal their type, the carrier can charge a higher (and therefore more appropriate) price for insurance for the high-damage millers.

The low-damage millers, however, may not be as likely to negotiate around the default rule as the high-damage millers. There are many more low-damage millers than high-damage millers. The costs of negotiating around the default rule may be the same for each miller, but the benefit that a low-damage miller receives from opting out of the default rule is smaller. By concentrating all the gains on the small class of millers with high consequential damages, the foreseeable damages rule creates a large incentive for them to come forward and negotiate around the default rule. A foreseeable damages rule might be preferable even if low-damage millers would opt out of a default rule that provided for consequential damages. A high-damage and a low-damage miller might each spend the same amount renegotiating a contract, but the total amount the low-damage millers spend renegotiating contracts is larger. There are more low-damage millers; hence, the transaction costs of negotiating around the default rule are higher.

Default rules can affect whether relevant information is disclosed when parties enter into their contracts. In a world in which each party needs information about the other in order to take the right amount of care, one cannot pick default rules solely on the idea that they should provide the parties with the terms for which they would have bargained if they were fully informed. Parties who are fully informed have incentives that are different from those who are not. The carrier can act optimally only if it is fully informed.

A consequential damages rule gives the carrier optimal incentives only if it can distinguish between high- and low-damage millers. If transaction costs are high and no one opts out of the default rule, the carrier will take too little care with respect to the high-damage millers and too much care with respect to the low-damage millers. Similarly, the carrier will take too little care with respect to the crankshaft of the high-damage miller in a foreseeable damages regime. When transaction costs are very low, parties will opt out of both legal regimes, and the default rule may not matter much.

Default rules matter most when transaction costs fall into an intermediate range. When we evaluate the two rules in this context, we have to examine their effects on the revelation of information. Under a consequential damages rule, the benefit to each low-damage miller from opting out may be too small for the miller to opt out of the default rule, given the costs of writing a special term for each miller. By contrast, the relatively small number of high-damage millers may be able to opt out of a foreseeable damages rule because the benefits to each miller are much larger and the costs of writing a special term for each miller are the same. Once high-damage millers engage in such bargaining, the carrier learns the type of miller and therefore how much care to take with respect to each.

Choosing default rules to take account of information problems becomes considerably more complicated when each party to the contract possesses information that is valuable to the other. *Peevyhouse v. Garland Coal and Mining Co.* is a good illustration.[18] Garland approached the Peevyhouses and bargained for the right to strip mine on their farm.

In return, Garland promised both to pay the amount specified in the contract and to restore the land when the job was completed. Garland subsequently broke its promise and refused to restore the land. The court, however, found that the cost of restoring the land exceeded the value the land would have after it was restored. The court held that the Peevyhouses were entitled only to the difference between the market value of the land if it were restored and its value in its unrestored state.

This case has been roundly attacked on the ground that the court failed to recognize that the value of the land to the Peevyhouses might exceed its market value. Anglo-American contract damages are built around the idea of putting the injured party in the position that party would have been in had the contract not been broken. If the land is worth restoring, the Peevyhouses should be given the cost of restoration. If it is not, they should receive an amount equal to the difference between the subjective value of the land to them if it were restored and the value of the land in its current state.

The appropriate default rule, however, should take account of the information problems the parties faced at the time they entered into the contract. Both sides may have possessed information that the other did not. Garland possessed private information about the probability of default and the costs of restoring the land, whereas the Peevyhouses possessed private information about the subjective value of the land to them. If transaction costs had been low, the Peevyhouses would have found out more about the likelihood of breach, and the parties could have bargained for a clause that would set out what damages Garland owed the Peevyhouses in the event of breach. Because transaction costs may be significant in this case, we need to ask how different default rules might have led the parties to draft a contract that would have been more tailored to their needs.

One can argue that the default rule should focus on the private, non-verifiable information that the Peevyhouses possess about the subjective value of the land. The situation is complicated, however, because there is a second information problem in this case.[19] At the outset, the Peevyhouses do not know the likelihood that Garland will breach as well as Garland itself knows it. This information may be verifiable information. Garland might, for example, be able to prove that it was likely to perform by showing that it had never broken any of its previous promises to restore land. Alternatively, the information might be nonverifiable. Garland can show promises it kept, but it is hard for the Peevyhouses to learn about the promises that it has broken. In addition,

Garland might not have faced similar situations in the past and its willingness to perform might be private, nonverifiable information. We need to consider both possibilities.

Assume for a moment that the information is verifiable information and unraveling forces Garland to disclose the likelihood that it will breach. At this point, we need consider only the private information that the Peevyhouses possess. Let us also assume that most people in the Peevyhouses' position value their land only at its market price. A few people, however, place a high subjective value on their land. Under these assumptions, a penalty default that awards only market damages in the event of breach might make sense. Those who do place a subjective value on the land would step forward and identify themselves. Garland could then decide whether to enter into the contract and how much to pay for its right to strip mine the land. Moreover, the damages clause for which the parties bargain would relieve the court of the impossible job of trying to determine the subjective value of the land, information that by its nature is private and nonverifiable. A penalty default may have the effect of requiring the Peevyhouses to take actions in the course of negotiating the contract that reveal the value they attach to the land.[20]

We should, however, also consider the possibility that the Peevyhouses have no way of learning directly about the likelihood that Garland will perform. The information may be private and nonverifiable because Garland may not have encountered similar situations in the past.[21] In this case, one could argue that it was more important to have a penalty default that gave Garland an incentive to disclose its private, nonverifiable information. To be sure, a default rule that gives the Peevyhouses the right to force Garland to restore the land in this case might lead to a waste of resources and might allow them a recovery in excess of the actual harm they suffer from the breach.[22] Even so, such a default rule might induce Garland to disclose information to the Peevyhouses. Once Garland tries to negotiate around the default, Garland may be forced to disclose information about the chance that the land will not be worth restoring. In addition, one may wish as a general matter to choose default rules that are detrimental to informed, sophisticated parties. Such parties know to bargain around them, and when they do so they may reveal information to their contracting opposites.

The ability of parties to add penalty clauses to a contract may give them the ability to signal information about their type that is otherwise nonverifiable. The common law rule banning penalty clauses may limit the ability of parties to signal nonverifiable information and therefore may not make sense, at least not between parties who are sophisticated commercial actors.

Screening and the Role of Legal Rules

The games in this chapter have been ones in which players who possessed nonverifiable information could signal their types by the actions they took. In other situations, however, the appropriate models are ones in which the uninformed player chooses actions that induce informed players to reveal their type. Such screening models also show a potential role for legal rules. The regulation of insurance contracts provides the best-known problems of this type.

An insurer faces two different kinds of information problems. First, an insurer must contend with *adverse selection,* or a problem of hidden information. Individuals have a better sense than the insurer of how likely they are to make a claim under a policy. At any particular price, the buyers of insurance who are the most likely to make a claim are also the most likely to buy the policy. If the insurer raises the price of the insurance in response, those who stop buying the insurance are the least likely to make a claim. This raises the price of insurance further still. Some types of risk may be uninsurable because of this adverse selection problem. The insurer increases the price because those at risk are most likely to buy it, but the increase in price makes the pool of those who buy the insurance riskier still. An equilibrium might not exist in which the insurer breaks even and someone is still willing to pay the price that the insurer must charge.

Insurance contracts must also take into account a *moral hazard problem,* a problem of hidden action. An insurer is not able to learn exactly how an individual behaves once the insurance contract is purchased. Hence, the contract cannot be written in a way that protects the insurance company from individuals taking actions that they would not take if they did not have insurance. Moral hazard problems can also be large enough to keep some kinds of insurance policies from being sold.

In many cases, adverse selection problems and moral hazard problems arise simultaneously. Both problems, for example, explain why no insurance company will sell a policy to cover a person's gambling losses in Las Vegas. There is an adverse selection problem because those who want such a policy and are willing to pay its premium are the ones most likely to have large losses. There is also a moral hazard problem. Once you have a policy that insures your gambling losses, you will gamble more recklessly than you otherwise would.

The mere presence of adverse selection and moral hazard problems does not in itself keep insurance markets from existing. Individuals are more likely to be careless when their house is insured and arsonists are especially eager to buy fire insurance, but a thriving market exists for such insurance. Nevertheless, legal regulations may make it possible for insurance markets to exist where they might not otherwise.

An insurer can respond to the problem of adverse selection by offering contracts that separate potential buyers of insurance by type. Let us assume that all buyers of insurance are risk averse and that an insurer is risk neutral and earns a competitive return. Buyers want insurance because insurance can equalize their wealth in good and bad states of the world. Someone who is risk averse prefers wealth of $50 in all states of the world to having an equal chance of $100 in one state and $0 in the other.

Let us assume that there are two types of insurance buyers. Half are high-risk and face a 50–50 chance of an accident that would leave them with nothing. The other ½ are low risk. They face only a 25 percent chance of an accident. A buyer's type is nonverifiable information. If everyone in the population were high-risk, the insurer could charge $50 for insurance and everyone would enjoy $50 in all states of the world. If everyone in the population were low-risk, the insurer could charge $25 and leave everyone with $75 in all states of the world.[23]

When both types exist, however, the insurer does not offer a single policy to both. The insurer is better off offering two policies, one for which it charges $50 and promises a payoff of $50, and another for which it charges less, but offers less coverage. The high-risk buyers would rather pay more for insurance and be completely covered than pay less and be only partially covered. The low-risk buyers, however, would prefer to live with some risk of being incompletely protected in bad states of the world rather than pay for complete insurance at the high price that the insurance company is charging for it. It costs low-risk buyers less to forgo complete insurance than to buy complete insurance, given that if they bought complete insurance they would be indistinguishable from high-risk buyers.

There is no guarantee, however, that the set of contracts that separates the two types of insurance buyers is sustainable either. Assume, for example, that an insurer who wants to offer insurance must announce the kind of policy or policies it will sell and cannot later withdraw or revise them. In such a case, the insurer will never offer a pooling contract. The insurer might be able to offer a contract that separates the buyers, but it is also possible that no equilibrium exists at all. There may be no contract or set of contracts that the insurer would offer and that the buyers would accept from which one or the other would not want to deviate. One can also build models with different assumptions about the insurer's ability to offer new contracts or withdraw old ones in which separating equilibria reemerge.

A legal rule that requires buyers to have insurance and insurers to offer a single pooling contract may cut the Gordian knot and solve the adverse selection problem. If employers cannot take health into account in hiring workers or determining what to pay them, requiring all employers to provide health insurance might produce a similar effect. The low-risk insureds may themselves be better off being pooled with the high-risk insureds than living in a world in which no insurance is available at all. The effect of the pooling contract is to transfer wealth from the low-risk insureds to the high-risk ones, but the social benefits to both types of insureds may justify the transfer.

Screening models may provide a way to understand how different legal rules affect the insurance markets and other environments, such as credit markets, in which the uninformed player tries to screen players by offering different kinds of contracts. For example, the standard contract between a small investor and a stockbroker mandates industry-run arbitration if an investor has a dispute with a broker. A broker could offer a contract that leads to information revelation among investors. Once the broker can distinguish among different investors, harmful discrimination may be possible. Let us assume that there are two types of investors: those who monitor brokers vigilantly and those who do not. Those who monitor brokers provide a benefit to those who do not. Because a few monitor, the others do not have to. The arbitration mechanism in the contract is unfavorable (we shall assume) to the investors, and the investors who monitor will refuse to enter into such contracts. The brokers will not modify the contract for them because objecting to the clause reveals that an individual monitors, and the broker would prefer not to do business with such a person. The arbitration provision, in other words, operates as a screening mechanism that allows brokers to attract only investors who are not vigilant.

Competition among brokers could lead some to offer a contract that attracts both kinds of investors. Investors who do not monitor should pay a premium to hire a broker who also has vigilant investors as clients. If competition among brokers had this effect, one might infer that arbitration clauses made sense after all. An equilibrium does exist in which all brokers offer a standard form contract providing terms that make every consumer better off. If the competitive equilibrium actually arises, an arbitration clause would exist only if it were efficient. Many equilibria could arise, however, that are not competitive. The problem we face is one that we encountered several times before: How do we identify which of several different equilibria are likely to emerge? This problem is particularly acute when parties interact with each other repeatedly over time, and it is this issue to which we turn in the next chapter.

Summary

In this chapter, we showed how a simple signaling game offered a way to understand the kinds of effects that legal rules might have in situations in which one party possesses private, nonverifiable information. In a pooling equilibrium, a player who possesses private, nonverifiable information that is unfavorable mimics the behavior of those who possess favorable information. Legal rules can change such inefficient outcomes in a number of ways. First, legal rules can limit the actions that informed players can take and shift the solution of the game from one pooling equilibrium to another, less inefficient pooling equilibrium. When all workers must be provided with orthopedic chairs, workers with bad backs receive the chairs that suit them, workers with good backs are only slightly worse off, and the employer still cannot distinguish between different types of workers and therefore still trains everyone.

Second, legal rules can limit the actions that the uninformed player can take in response to the informed player. When the uninformed player is disabled from acting on inferences drawn from the informed player's actions, it is possible for an efficient separating equilibrium to emerge. Even if the separating equilibrium that emerges is not efficient, it may be an unambiguous improvement over the pooling equilibrium because one of the types of informed player is better off and the uninformed player is no worse off. The rule takes away from the uninformed player only a right to take an action that would not be taken in equilibrium anyway. To understand the effects of a legal rule, we must pay attention to the way it affects not only actions that the parties actually take, but also actions that they would not have taken even in the absence of a legal rule. A player with unfavorable information loses the incentive to mimic the actions of the other informed player if the informed player cannot take advantage of any information that is inferred from the actions of the other players.

Third, legal rules can require uninformed players to pay damages unless they take certain actions. Such a rule may again bring about an efficient separating equilibrium. Assessing the effectiveness of this kind of legal rule—or, indeed, the effectiveness of introducing any legal rule at all—requires a sensitivity not only to the information that the parties possess, but also to the information available to the legislature when it passes a law and the information available to a court that must enforce it. Concerns about information, for example, may lead one to favor a damages rule in some cases. Because such a rule forces the uninformed player to internalize the costs and benefits of certain actions (such as offering training), the lawmaker who crafts such a rule may need less information about the costs and benefits of the action.

When many different contracts are being offered in the marketplace, informed players may find it in their interest to go to one or another, depending on the type of player they are. The existence of many different options may make it less likely that those with unfavorable information will mimic other players. Workers with bad backs or workers who expect to become parents may choose different kinds of jobs. Our basic model of signaling analyzes situations with a single employer and thus does not capture the possible sorting that might take place when there are many employers. Workers with bad backs might tend to look for jobs and employers that are different from those that workers with good backs want. Similarly, the greater the ability of parties to negotiate with each other, the more they may be able to devise mechanisms that transform nonverifiable information into information that is verifiable, provided that the information is observable to someone, if not to the parties or to a court. The problems we have confronted in this chapter, however, never disappear completely, and the forces that we have identified lie beneath the surface of many debates about law reform and the effects of different legal regimes.

Bibliographic Notes

Nonverifiable information and signaling. The literature on refinements of perfect Bayesian equilibria is large. The basic structure of the game we develop in this section is set out and the solution concepts used to solve it are explored in Cho and Kreps (1987). Cho and Kreps examine different solution concepts using a story involving two restaurant patrons (a tough guy and a wimp) who prefer beer and quiche for breakfast respectively, but who encounter a bully who enjoys fighting wimps but not tough guys. The breakfast order of the tough guy or the wimp turns on the inferences each believes that the bully will draw from the order. Cho and Kreps use this story to motivate refinements that eliminate seemingly implausible solutions to this game (in particular, the solution in which both the tough guy and the wimp order quiche). In the last several years, the *beer-quiche game* has joined the prisoner's dilemma as one of the central paradigms of game theory.

Other important papers on formal models of nonverifiable information include Banks and Sobel (1987), Kohlberg and Mertens (1986), Grossman and Perry (1986), and Farrell (1983). A textbook, such as Fudenberg and Tirole (1991a) or Myerson (1991), is the best place to go for a comprehensive discussion. Gibbons (1992) has clear, simple examples. Our discussion of centralized and decentralized rules was influenced by Farrell (1987) and Bolton and Farrell (1990) as well as by Kaplow (1992) and Posner (1992). The classic model of adverse selection is Akerlof's (1970) analysis of market breakdown in the market for used cars. Signaling models have been applied to a great number of economic settings. Spence (1974) is one of the earliest and contains the model of education as a signal of productivity in the labor market. Applications of signaling models to contract law issues include Aghion and Hermalin (1990), which also models the problem of parental leave legislation.

Information revelation and contract default rules. The information revelation aspects of contractual terms are based on the analysis of Ayres and Gertner (1989). They discuss the problem of both *Hadley v. Baxendale* and *Peevyhouse v. Garland Coal and Mining Co.* Other papers that focus on these issues include Johnston (1990), Bebchuk and Shavell (1991), Ayres and Gertner (1992), and Goetz and Scott (1985). Maute (1993) offers a comprehensive study of *Peevyhouse.*

Signaling, screening, and the costs of information. Screening models have also been applied to a great number of different settings. The classic paper on insurance is by Rothschild and Stiglitz (1976). Stole (1992) examines penalty damages as a screening problem.

CHAPTER 5

Reputation and Repeated Games

To this point, we have examined interactions between individuals as isolated events. Individuals, however, often have repeated dealings with one another. They must take account of how the decisions they make in one interaction will affect what happens in future ones. We can gain an understanding of how people are likely to behave in these situations by examining the strategies that rational players choose in games that consist of a simple game that is repeated many times. We use these models to examine a number of antitrust problems, including tacit collusion, conscious parallelism, and predatory pricing.

Before we look at repeated games, however, we return to the backwards induction solution concept. Backwards induction is useful for analyzing many games, but it rests on assumptions that may generate implausible results when applied to games that are repeated many times. Hence, we must be careful about applying it here. In the next section, we examine backwards induction and its limits more closely by applying it to two commercial transactions, one involving negotiations before a formal contract is signed and the other dealing with an installment sale.

Backwards Induction and Its Limits

Many contract rules establish governing principles for contract negotiations. A good example is the legal rule that conditions the legal enforceability of a contract on the existence of a signed writing evidencing the agreement of the parties. This rule, called the Statute of Frauds, applies to many contracts, including any contract for the sale of land and any contract involving the sale of goods worth more than $500.

The Statute of Frauds does not require that the terms of the contract be set out in any detail. The writing need only show that the parties have entered into a contract; it does not require anything even as rudimentary as a price term.

We can ask whether such a rule is desirable. To be sure, parties often find it in their interest to reduce their contracts to writing, but it might seem that the parties themselves are best positioned to decide whether a written contract brings benefits worth its costs. The rule does make it hard for people to assert that they have a contract with someone with whom they have never dealt. It is difficult, however, to justify the rule on this basis. The risk of liability that a complete stranger poses to a commercial actor seems a small one. After all, a complete stranger will not be able to persuade a finder of fact that a deal existed without being able to provide some evidence of a deal. In the absence of any credible evidence that the stranger ever met the person being sued, it is hard to see how such a lawsuit could be successful.

The Statute of Frauds, however, may reduce the risk of problems that can arise during the course of negotiations that typically take place before a contract is entered. The facts of *Southwest Engineering Co. v. Martin Tractor Co.* illustrate this kind of problem. Southwest was a general contractor that was installing runway lighting at an Air Force base. It needed to buy a stand-by generator and made inquiries with Martin Tractor. A series of negotiations eventually led to a lunch in an airport cafeteria. During this lunch, Martin's employee took out a piece of paper and listed the components of the generator as well as their price. Southwest's employee took a copy of this piece of paper.

The deal fell through a few weeks later and Martin attempted to withdraw "all verbal quotations." Southwest brought suit against Martin, alleging that they had reached an agreement during the lunch. Martin defended its position on the ground that, regardless of whether an agreement was made, it was not legally enforceable because the price quotation that its employee had written on the piece of paper was not sufficient to satisfy the Statute of Frauds. The court ultimately rejected this argument.

In commercial dealings such as the one in *Southwest Engineering*, a series of meetings between the buyer and the seller is commonplace. Parties may reach agreement on many terms, and other terms may be open. In a world in which there is no Statute of Frauds, the moment at which a legally enforceable contract is formed may be unclear. There may be several points during the negotiations at which one party or the other could insist that a deal of some kind had been reached even when (at least in the view of the other party) none had been reached. *Southwest Engineering* may be such a case. Southwest's employee may have thought that there was a deal, whereas Martin's employee may have thought that details needed to be worked out or that approval of the home office was still necessary.

Before parties sign a formal contract, they may negotiate with each other over an extended period of time. They may discuss possible deals orally. In other cases, a price term may be set down in writing. At every point, however, it may be uncertain whether a contract exists or what its terms are. Consider what might happen if no Statute of Frauds existed. If the parties stop short of signing a written contract, one or the other party may be able to assert that a binding contract nevertheless exists that has terms favorable to that particular party.

We can model this problem as a game in which the players must repeatedly decide whether to continue to negotiate or to exit from the negotiations. The payoff that a party enjoys after one or the other exits turns on whether the negotiations have progressed so far that one or the other can claim that a legally enforceable contract exists. Figure 5.1 presents the simplest version of such a game. Each party has a chance to exit during the negotiations and assert that a deal exists that is favorable to that party but unfavorable to the other. A player who exits enjoys a payoff of $5 but leaves the other with a payoff of −$5. If neither party exits and both sign a formal contract, each receives a payoff of $3.

We can solve the game in Figure 5.1 by using backwards induction. At the start of the game, the buyer must decide whether to ask for a

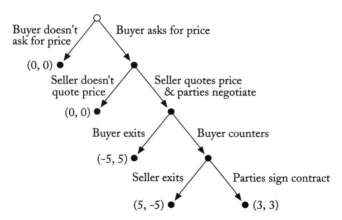

Figure 5.1 No writing requirement. *Payoffs:* Seller, Buyer.

price quote. The buyer cannot decide without forecasting the subsequent play of the game. One useful way to do this is for the buyer to look at how the seller behaves at the last node. At the last node, the seller's best response is to exit rather than to sign the contract. For this reason, the buyer's best response at the next-to-last node is to exit as well. Exit brings a payoff of $5, whereas continuing to negotiate brings the buyer a loss of $5, given that the seller exits if offered the chance. At the previous node, the seller's best response is again to exit. The seller prefers a payoff of $0 to a loss of $5. Because the buyer takes these best responses into account at the start of the game, the buyer is indifferent between never commencing negotiations and asking for a price quote, because in either case the buyer will receive a payoff of $0.

Serious negotiations are never undertaken because of the risk that one or both of the parties to the negotiation will opportunistically claim that a contract in fact exists. As a result, the parties cannot reach the best outcome—that in which a contract is signed and each party receives a payoff of $3. The Statute of Frauds, however, may solve this kind of problem. If a legal action can be brought only if a contract is signed, no player can gain an advantage from exiting. If the payoffs of (0, 0) and (0, 0) replace the payoffs of (−5, 5) and (5, −5), the parties will complete the contract and enter into a mutually beneficial trade.

Under this view, the Statute of Frauds may make parties more willing to begin negotiations with each other. If the Statute of Frauds is to serve this purpose, however, it must require more than evidence that negotiations took place. In *Southwest Engineering* and other Statute of Frauds cases, no one doubts that there were negotiations. A writing requirement, if we have one at all, should be strong enough to provide an answer to the question of whether the negotiations led anywhere. Because the writing in *Southwest Engineering* did not show any more than that the parties had met and negotiated, one can argue that the Statute of Frauds was not satisfied and that the case should have been decided differently.

The extensive form game in Figure 5.1 isolates a principle that may underlie the Statute of Frauds. This perspective may also help us to appreciate why debates over rules such as the Statute of Frauds are so central to the conceptual foundations of contract law. This analysis can also be extended in a straightforward way to other rules governing the formation of contracts, such as those that refuse to recognize agreements that are too indefinite. Unlike warranty and other gap-filling terms, rules such as these set the climate in which negotiations take place. It is hard for parties to opt out of them, and the stakes involved in getting them right are much higher.

Backwards induction has an intuitive appeal and is easy to apply in many games. Nevertheless, it may lead to solutions that seem implausible. Consider the following game. Seller and Buyer contemplate entering into an agreement. Seller would promise to ship Buyer goods in 100 equal installments and Buyer would promise to pay for each installment after receiving it. Seller has no recourse against Buyer if Buyer refuses to pay, perhaps because legal remedies are too costly. (Seller's only option is to cease any further shipments.) Similarly, Buyer has no recourse if Seller chooses not to perform. It costs Seller $1 to make each installment. Buyer pays $2 for each shipment of goods and the goods have a value to Buyer of $3. The game between Seller and Buyer takes the extensive form game illustrated in Figure 5.2.

To use backwards induction to discover the likely course of play, we

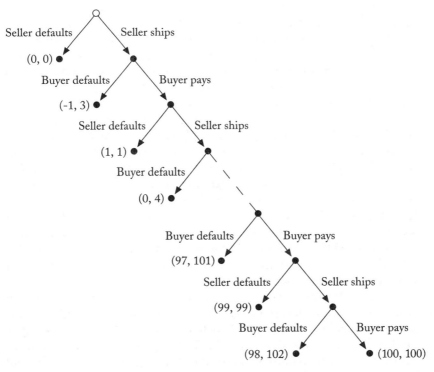

Figure 5.2 Installment sale. *Payoffs:* Seller, Buyer.

again start at the end of the game, with Buyer's decision after it has received the last installment. It can either refuse to pay for the goods and enjoy a payoff of $102 or pay $2 for the last installment and enjoy a payoff of only $100. Buyer is thus better off if it does not pay. Seller anticipates that Buyer will not pay for the last installment. On its last move, Seller therefore faces a payoff of $98 if it ships the goods, but $99 if it does not. Back one more step, Buyer recognizes that Seller will not ship the last installment of goods and Buyer therefore will not pay for the second-to-last shipment of goods. (Buyer enjoys a payoff of $101 if it defaults, instead of a payoff of only $99 if it pays for the goods.) Seller makes a similar calculation on its next-to-last move, and the process continues until we reach Seller's first move—in which it decides not to ship at all. In short, backwards induction suggests that the entire deal will unravel because each party will anticipate that the other will default on its next move.

Common sense tells us that applying backwards induction to this model does not capture how Buyer and Seller would deal with each other. Buyer and Seller might enter into this transaction because some force is at work that we have not taken into account. Buyer, for example, may have trouble buying goods from others if it failed to pay Seller. Moreover, both parties might keep their promise because of the prospect of future dealings with each other. Even apart from reputational issues, however, backwards induction seems to do a poor job of predicting how the game itself might be played.

This game is one that involves many iterations and in which there is only a small cost to a player of deviating from the equilibrium strategy of defaulting at every opportunity. For this reason, the game places unusual weight on the assumption that the payoffs that the parties enjoy are common knowledge. We assume that Buyer and Seller are certain that the game has the payoffs that we show, that each knows that the other player is certain, that each knows that the other player knows, and so forth. They know that if the other did not default and either shipped the goods or paid for them, it would be due to a mistake, a *tremble*, rather than an intentional decision to adopt something other than the equilibrium strategy.

Consider, however, how the solution to the game might change if either player was not certain that a deviation from the equilibrium was simply a mistake that was not likely to be repeated. Focus, for example, on Buyer's move after it receives the first shipment from Seller. It must decide whether to default on its payment obligation and enjoy a benefit of $3 or pay and risk that Seller will default on the next move and leave it with only $1. If Buyer is not certain that Seller has made a mistake, it must consider how it should respond given that Seller has already acted in a way contrary to the predicted equilibrium.

If Buyer is only slightly unsure about Seller's payoffs, Buyer's best response in a game such as this may be to experiment and cooperate for an additional period. Buyer has little to lose if Seller turns out to have made a mistake, but it has a great deal to gain if Seller's payoffs are in fact different from what they appear to be. Once Buyer is willing to experiment by continuing the game for an additional period, however, Seller has something to gain from making the first installment. Even if Seller's payoffs are exactly what Buyer thinks, Seller may still be better off shipping the goods as long as Buyer harbors some doubts.

We return at the end of the chapter to the question of how uncertainty in the payoffs can affect the play of repeated games. At the moment, however, we want to sound a note of caution. The assumption of common knowledge is a strong one. We must be careful about using it, particularly in games with many iterations and in which deviations from the equilibrium come at low cost and may bring substantial benefits relative to the equilibrium outcome. The test of a concept such as backwards induction is ultimately how well it captures the way people actually behave. Although this solution concept is quite powerful, one cannot apply it or any other solution concept blindly.

Infinitely Repeated Games, Tacit Collusion, and Folk Theorems

When parties interact with each other over time, they often do not know when the interactions will end. The solution to the installment sale game in Figure 5.2 turned on the game's having a definite terminus. In this section, we want to examine games in which there is no definite end. Manufacturers in a small market can earn supracompetitive profits if they simultaneously raise prices and reduce output. The antitrust laws ban explicit cartel agreements, but less clear is how they affect tacit agreements. Tacit agreements may exist when firms in a market increase prices and reduce output without actually entering into an explicit bargain. To understand this issue, we must first examine how tacit agreements could ever come into being and survive for any length of time. A useful way of doing this is to model the pricing and output decisions of those in an industry with few firms as a repeated game of indeterminate length.

In the 1940s, the leading cigarette manufacturers were accused of violations of antitrust laws in the pricing of cigarettes.[1] Pricing in the industry was identical across firms. When one firm changed prices, competitors immediately matched them. The government argued that this pattern allowed the firms to sustain prices at noncompetitive levels. The Supreme Court upheld the jury verdict against the tobacco manufacturers, even though the suit did not allege that there had been any explicit agreement or conspiracy to fix prices.

Subsequent cases, however, have made it clear that the plaintiff in an antitrust case must do more than show parallel behavior among the major firms. Indeed, even conscious parallel behavior falls outside the reach of the antitrust laws. For conduct to run afoul of antitrust laws, some additional "plus factors" must also be at work. In this section, we explore how firms can cooperate on price without explicit agreement and how a variety of practices may facilitate their ability to collude tacitly. These in turn cast light on what one should have to show beyond parallel behavior in order to justify taking action against particular firms.

All models of tacit collusion rely on both the dynamic nature of price competition and the ability of firms to respond to their rivals' pricing decisions. To develop the ideas formally in a simple model, let us begin by assuming that there are two firms in the industry and there are only two prices they can charge: High or Low. In each period they simultaneously decide which price to charge. Profits in that period are given in the bimatrix shown in Figure 5.3.

Joint profits are greatest if both firms choose a high price. If the game is played only once, however, charging a low price is a dominant strategy for each player. If the second firm charges a high price, the first firm earns $6 more by charging a low price rather than a high price. If the second firm charges a low price, the first firm earns $5 more by charging a low price rather than a high price. Thus, no matter what the second firm does, the first firm earns more by charging a low price. Because the game is symmetric, the second firm must also find it in its interest to charge a low price.

		Firm 2	
		High	Low
Firm 1	High	10, 10	0, 16
	Low	16, 0	5, 5

Figure 5.3 Tacit collusion. *Payoffs:* Firm 1, Firm 2.

This game has the same structure as the prisoner's dilemma. Both players adopt the low-price strategy, even though they would be jointly better off if they adopted the high-price strategy. A game in which the firms could choose their price from a continuum is also a game where the unique equilibrium is one in which both firms charge a low price. (The equilibrium is Nash, however, rather than one in which each player has a strategy that is strictly dominant. For any given price above marginal cost, the best response is to undercut it by only a small amount.)

Let us now imagine that the game is repeated a finite number of times. We encounter the same problem we saw in our discussion of the installment sale game represented in Figure 5.2. Each player posits the play of the last period. In the last period, neither has any concern about how an action in this period will affect future play because, by definition, the game will end after this period. In the last period, both act as if they are playing a one-shot game. Each will therefore play its dominant strategy and charge the low price.

Given the outcome in the last period, both players will find it in their interest to charge a low price in the next-to-last period as well. Each player knows that the other will charge a low price in the last period no matter what happens in this game. Hence, both will play in the next-to-last period as if it were a one-shot game as well. They will engage in the same process in the period before the next-to-last one also. The process repeats itself until the players reach the first period. Backwards induction suggests that the players will charge the low price in every period. Mere repetition of the strategic interaction does nothing to allow cooperative pricing.

As in the installment sale game, backwards induction predicts this course of play because of strong assumptions of common knowledge. In this case, the equilibrium turns crucially on the number of repetitions of the game being common knowledge. This assumption is unrealistic. There is no set amount of times that firms in an industry compete with each other, and our model should take this into account.

The finitely repeated game is the appropriate model only if the parties know with certainty exactly when the game will end. For this reason, we focus in this section on infinitely repeated games. A game of infinite length seems equally unrealistic, but, as we shall show, a game of infinite length and a game of uncertain length with a fixed probability of ending after each period have the same structure.

Let us assume that the two firms play the game in Figure 5.3 an infinite number of times. A game that consists of an infinite number of repetitions of another game is a *supergame*. Once we move to infinitely repeated games, we need to understand the relation between payoffs in one game and payoffs in later ones. If a payoff of a fixed amount in the future were worth as much to a player as the same amount today, a player would be indifferent between a strategy which leads to a payoff of $10 each period and a strategy which leads to a payoff of $5 each period. Each strategy yields payoffs that, when summed, are arbitrarily large.

We shall assume that a player values a payoff in a future period less than a payoff in the present period. We account for this by introducing a discount factor, the amount by which the value of a payoff in the next period must be adjusted to reflect its value in the present period. If we have a discount factor of δ, the present value of one dollar earned in the subsequent period is δ dollars.[2]

We can now compare this game to one whose duration is uncertain. The uncertainty could be due to a possible exit by one firm, introduction of a superior product by a competitor, government regulation, or a variety of other factors. Let us assume that the firms believe that the probability that the game will end in any period is $1 - d$, so the probability that it will continue is d. When deciding on a strategy in period 1, a firm discounts the next period's profits by δ times the probability that the game will continue another period, which is d. The firm discounts profits two periods in the future by δ^2 times the probability that it will still be playing, which is d^2.

When we model games of uncertain length, we can use the product of δ and d in every term. More simply, we can continue to use δ, but interpret it as the product of the probability that a game will end in a given period and the amount by which payoffs in each subsequent period are discounted. A game with an uncertain end is equivalent to the infinitely repeated game with a discount factor that is again discounted by a factor equal to the probability that the game will continue in any period. The infinitely repeated game is appropriate as long as there is some uncertainty about when the game will end.

Because there is no last period, we do not have a definite end point from which to start the process of backwards induction. To solve these games, one must describe different strategies and see if any combination of them forms a subgame perfect Nash equilibrium. For example, the strategies "charge a high price in even periods and a low price in odd periods" and "charge a high price in the first period and, in subsequent periods, copy what the other player did on its previous move" are both fully specified strategies in the sense that they tell us how a player moves at every information set in the game. We can then ask whether such strategies are subgame perfect by looking at all the possible subgames that could arise both on and off the equilibrium path.

Consider first the following *grim* (or *trigger*) *strategy:* Both firms adopt the strategy of charging a high price in the first period. In all subsequent periods, the firms charge a low price if either firm has ever charged a low price in any previous period, otherwise they charge a high price. First, we will confirm that each firm choosing this strategy forms a Nash equilibrium if the discount factor δ is sufficiently high. Let us consider what Firm 1 gets if it plays its equilibrium strategy. If both firms play their equilibrium strategies, they will each end up charging the high price in each period, since neither initiates a price war. Thus, the present value of profits for Firm 1 is $10/(1 - \delta)$.

At the beginning of each period, the present value of expected profits is equal. The continuation game at the beginning of the tenth period looks identical to the game at the beginning of the first. This game, in other words, is *stationary.* The strategy space of a player, the choices available to a player at any point in the game in which that player moves, remains the same. This stationary structure simplifies the analysis of infinitely repeated games.

We need to ask what will happen if Firm 1 deviates from its strategy and decides to charge a low price at some point, even though both players have charged a high price in all preceding periods. The profit from charging a low price this period is $16. In all subsequent periods, however, Firm 1 will receive $5. (For both firms to charge a low price from then on is a subgame perfect equilibrium.) Thus, the returns from charging a low price are $16 + 5\delta/(1 - \delta)$. Charging a high price when the other player adopts a grim, or trigger, strategy is therefore more profitable than charging a low price as long as

$$\frac{10}{1 - \delta} > 16 + \frac{5\delta}{1 - \delta}.$$

By solving this inequality, we learn that charging a high price is better than charging a low price as long as $\delta > {}^6\!/{}_{11}$. A player receives higher profits today by charging a low price, but the other player responds by charging a low price in all subsequent periods. This punishment brings lower profits in the future that may offset the short-term gains. A player has less incentive to charge a low price as δ rises. Indeed, as δ rises, a player cares more about the profits in future rounds relative to profits in the current round. If $\delta > ^6/_{11}$, the strategy combination in which the players both charge a high price until the other charges a low price forms a Nash equilibrium.

Given that Firm 2 is playing this strategy, Firm 1 earns $10/(1 - \delta)$ if it plays its equilibrium strategy of charging a high price as long as Firm 2 has never charged a low price. We have just shown that changing to a strategy in which it charges a low price leaves Firm 1 worse off when $\delta > ^6/_{11}$. The only other strategies that Firm 1 could adopt would be ones in which it did not initiate a low price. (Such a strategy would include, for example, one in which a player charges a high price until the other charges a low price and, in that event, charges a low price for five periods and then returns to the high price.) These strategies, however, also yield $10/(1 - \delta)$, given Firm 2's strategy of charging a high price unless Firm 1 charges a low price. All other strategies in which Firm 1 continues to charge a high price offer the same payoff as a trigger strategy. For this reason, a trigger strategy is a best response to the trigger strategy of Firm 2. Hence, a Nash equilibrium exists when both players adopt it.

This strategy combination is also a subgame perfect Nash equilibrium. We can see this by examining the possible subgames and making sure that the strategies form a Nash equilibrium in each of them. There are an infinite number of subgames (one for each possible history at each period), and we cannot check them one by one, but we can nevertheless examine the full spectrum of possible situations. Consider any subgame in which no firm has yet charged a low price. The argument of the preceding paragraph showed that the strategies form a Nash equilibrium in these subgames.

Next consider the subgames in which Firm 1 has charged a low price in some previous period. Firm 2's strategy is to charge a low price in all future periods. From this point on, Firm 2's strategy is independent of Firm 1's actions (it will always charge a low price), so Firm 1 should maximize current profits in each period. In each period, charging a low price is a dominant strategy for Firm 1. The same reasoning applies for Firm 2. Neither has reason to charge a high price when the other is going to charge a low price. Because each firm's strategy is a best response to its rival in these subgames, the strategies form a Nash equilibrium. Having covered all subgames, we have established subgame perfection.

The intuition that explains why these strategies form an equilibrium is simple. If both firms charge a high price initially, but charge a low price forever if the other ever charges a low price, neither player does better by charging a low price. The short-term gains are insufficient to compensate for the future losses from an infinitely long price war. Given that the best response to a low price is always a low price, a Nash equilibrium exists when both firms charge a low price.

The trigger strategy is not the only equilibrium in the infinitely repeated game. As suggested, for example, it would also be a subgame perfect Nash equilibrium if each firm charged a low price in every period. Neither firm can increase its profits in any period from deviating. A high price lowers current profits and has no effect on future profits. In fact, as we shall soon see, there are an infinite number of subgame perfect Nash equilibria. When δ is sufficiently high, one such strategy is a version of *tit-for-tat*. A firm playing the strict tit-for-tat strategy begins by charging a high price and then charging a low price only if its rival charged a low price in the preceding period. If both players adopt tit-for-tat, each begins with a high price and charges a high price in every period.

Tit-for-tat stands in sharp contrast to trigger strategies in which a single transgression is punished in all subsequent rounds. If a firm deviates from tit-for-tat against a rival playing tit-for-tat, it goes back to a cooperative price, and the punisher returns to the cooperative price in the subsequent period. The firm that deviates suffers one period in which it charges a high price and earns nothing because the other player punishes by charging a low price. In subsequent periods, however, both parties can return to charging a high price. A single defection need not lead to a low payoff in all subsequent periods.

A Nash equilibrium of the supergame exists in which both players adopt tit-for-tat if δ is sufficiently high. There are two types of deviations to consider. First, a firm can charge a low price and never revert to a high price. This firm will enjoy high profits for a single period and then $5 forever. This is equivalent to the payoffs we analyzed earlier. Hence, all that is necessary to prevent this deviation is to ensure that $\delta > 6/11$.

Second, instead of playing the equilibrium strategy, a firm can charge a low price for a single period and then revert to a high price. If Firm 1 follows this strategy, it charges a low price in the first period and Firm 2 charges a high price. In the second period, Firm 1 charges a high price and Firm 2 charges a low price as punishment. The third period is just like the first period because Firm 1 knows that Firm 2 will charge a high price. Thus, if the deviation was profitable in period 1, it will also be profitable in period 3. The game will cycle. Firm 1's profits from following this strategy are $16 in odd periods and $0 in even periods. The discounted value of this deviation is $16/(1 - \delta^2)$. A player is better off playing the equilibrium strategy than this one as long as

$$\frac{10}{1 - \delta} > \frac{16}{1 - \delta^2}.$$

Hence, the equilibrium strategy is more profitable than alternating between high and low prices as long as $\delta > \frac{6}{10}$. When δ is greater than $\frac{6}{10}$ (and thus sufficiently large to make deviation to alternating between high and low unattractive), a player's best response is to charge a high price in every period when the other player has adopted tit-for-tat. There are some values of δ (for example, those in which it is larger than $\frac{6}{11}$, but smaller than $\frac{6}{10}$) in which tit-for-tat is not an equilibrium even though infinite punishments for cheating are. For many values of δ, however, both strategies generate an equilibrium when each player adopts them.

We need a slightly different version of tit-for-tat if the strategy combination is to be subgame perfect. Each firm starts with a high price and charges a high price if both cooperated in the previous period or if both defected; otherwise each charges a low price. This formulation of tit-for-tat is necessary to ensure that there is cooperation rather than constant cycling in some subgames that are off the equilibrium path, in particular subgames that begin after one of the players has defected in the previous period.

So far, we have set out three symmetric subgame perfect Nash equilibria of the infinitely repeated game. Indeed, if the discount factor is sufficiently high, any pattern of high prices and low prices can be sustained with the threat to revert to low prices forever if anyone deviates from the pattern. This result can be generalized to all supergames with complete information. That is, any payoff that gives each player at least what that player receives in the single period Nash equilibrium forever, and that is feasible given the payoffs in the game, can be supported as a subgame perfect Nash equilibrium of the supergame. In Figure 5.3, any combination of payoffs that allows each player to average at least $5 in each round is part of some subgame perfect Nash equilibrium. This result and others like it are known as *folk theorems* because they were part of the received lore of game theory before they appeared in any published papers.

In many repeated games, there are multiple subgame perfect Nash equilibria, and we again face the familiar problem of choosing among them. We might first consider whether players would be likely to adopt the subgame perfect equilibrium that was *Pareto-optimal*. One might argue that, if one equilibrium gives each party a higher payoff than another, rational parties should never play the second equilibrium. If we further focus on symmetric equilibria, there will be unique predicted payoffs in the game, which may be supported by many equilibrium strategies.

This prediction, however, seems suspect. Pareto-optimal equilibria rely on very severe punishments. The greater the punishment from deviating, the more likely parties are to cooperate. These equilibria are plausible, however, only if all the players believe that the punishments are in fact carried out when someone deviates. The threat of the severe punishment needed to support the Pareto-optimal equilibria may not be credible. Parties may not find it in their self-interest to carry out severe punishments after there has been a deviation.

If parties always want the Pareto-optimal solution, they should adopt the Pareto-optimal equilibrium of the subgame that arises after the deviation. This Pareto-optimal equilibrium, however, would not take into account the need to punish the previous deviation. It would likely provide a less severe level of punishment than is necessary to support the original Pareto-optimal equilibrium. One can argue that parties are likely to adopt a Pareto-optimal equilibrium in the game as a whole only if one can explain why these same parties do not adopt strategies in each subgame that are also Pareto-optimal. If parties consistently gravitate toward the Pareto-optimal equilibrium, they should do it in subgames as well as in the game as a whole.

Another approach is to identify equilibria that are *renegotiation-proof*. Such an equilibrium is one in which the nondeviating party still has the incentive to carry out the punishment should it prove necessary. When the nondeviating party punishes the other party for deviating, the nondeviating party must be willing to incur the costs of inflicting punishment. This way of isolating the strategies that players are likely to adopt raises complications, however. In the subgames that follow any deviation from the equilibrium, the deviating party must willingly choose strategies that yield payoffs that are sufficiently low to make the original deviation unprofitable. One needs to explain why a player who deviates would also be willing to accept low payoffs so that both players can return to the equilibrium path.

None of the existing methods of narrowing the possible equilibria proves completely satisfactory. Nevertheless, the models do show that repetition itself creates the possibility of cooperative behavior. The mechanism that supports cooperation is the threat of future noncooperation for deviations from cooperation. One of the crucial determinants is the discount factor. The higher the discount factor, the more players have to gain from cooperation. If firms can change prices frequently, the appropriate period length is very short and the corresponding discount factor is very high. If firms can change prices only infrequently, as in catalogue sales, the appropriate discount factor will be much lower. What matters, of course, is not whether firms actually change their prices, but whether they have the ability to do so.

When firms in an industry seem to change their prices in unison, we should pay attention to how often prices change. Tacit collusion leading to noncompetitive pricing is much more likely in industries in which firms can adjust prices almost instantly (as they can, for example, in the airline industry) than in other industries (such as the mail-order business) in which they cannot. Firms in the latter industry are more likely to be responding to changes in supply and demand than to a tacit agreement among themselves.

One of the major impediments to tacit collusion is the inability to observe the prices charged by rivals. In many markets, prices are negotiated between buyers and sellers in a way that makes it impossible for rivals to observe each other's prices directly. How does a firm know when a rival has undercut the collusive price? Eventually one may figure it out by a drop in demand or by information from potential customers. If it takes a long time to learn, then the appropriate period length is quite long, and collusion may be difficult to support. Mistakes are also possible, as a firm's demand could drop for reasons other than a rival's secret price cut. In such environments, strategies that lead to cooperation may fare relatively poorly.

Consider what would happen if players were to adopt tit-for-tat when there was uncertainty about the actions of the other players. Assume that a player erroneously believes that the other player defected. The other player may see that player's decision to punish as a defection that triggers punishment in return. Tit-for-tat requires the first player to respond noncooperatively to the second player's response, which in turn generates a noncooperative response from the second player. Even though both players have adopted tit-for-tat, a single mistake could lead them to act noncooperatively forever, or until another mistake is made and a party mistakes punishment for cooperation. Other equilibria that support cooperation suffer from similar problems. For this reason, one might suspect that tacit collusion is more likely in industries in which pricing information is readily available. A good example is the airline industry, in which price schedules are publicly distributed over a computer network to travel agents and others.

As the number of firms in an industry increases, collusion becomes less likely. By deviating to a lower price and expanding output, one may have a fairly small impact on each of a large number of firms. It becomes less likely that such firms will have an incentive to punish. The detection of cheating also becomes significantly more difficult as the number of firms increases. A firm with many competitors may experience a decrease in market share and have much more trouble determining whether one firm cheated or simply became more efficient.

Even if the other firms can tell that one firm has cheated, it will be hard for them to coordinate the appropriate period of punishment. It might not be in the interest of all the firms to punish the deviating firm by charging a low price, but there may be no way for them to agree on which firm should act as the enforcer. The Fourth Circuit in *Liggett Group, Inc. v. Brown and Williamson Tobacco Corp.*[3] noted exactly this problem:

> Oligopolists might indeed all share an interest in letting one among them discipline another for breaking step and might all be aware that all share this interest. One would conclude, however, that this shared interest would not itself be enough . . . The oligopolist on the sidelines would need to be certain at least that it could trust the discipliner not to expand the low-price segment itself during the fight or after its success.

Changes in the collusive price over time may limit the ability of firms to collude. Let us say that a firm receives private information about changes in demand or cost conditions that imply that the collusive price has dropped. If the firm lowers its price, rivals may think this is an attempt to cheat and respond with punishment. The result may be that firms are unable to change prices as often as is optimal or price wars may break out through mistaken punishments.

One practice that can solve this problem is price leadership, in which price changes are initiated by a single firm in the industry and are followed by all other firms. Differences among firms may also make collusion more difficult, as one firm may wish to cut prices because of lower costs or different demand for its product. Firms will not agree on a single optimal price for the industry, and each firm may compete to induce rivals to accept its preferred price. All of these results have again been captured in models that have the same basic structure as that in Figure 5.3.

Models such as this one suggest that antitrust enforcers pay attention to how rapidly prices change in an industry, how information about prices is conveyed, and how many players there are. We can ask if existing law shares the same focus by looking at the rules governing facilitating devices—industry practices that are thought to make collusion more likely. Specifically, we can look at the scrutiny that antitrust law gives to trade associations. A trade association often collects information about prices, costs, and demand from the member firms. This information may make the detection of cheating a great deal easier. In an oligopoly, resources devoted to monitoring competitors' behavior are a *public good*. The benefits of monitoring accrue to all firms. Thus, individual firms may try to *free ride* on their competitors' monitoring efforts, and less monitoring takes place than if firms could reach explicit agreement. A trade association can get around this problem.

The jurisprudence of trade associations and sharing information among competitors dates to 2 cases in the lumber industry from the 1920s. *American Column and Lumber Co. v. United States* concerned the American Hardwood Manufacturers' Association, a trade association of 400 U.S. hardwood manufacturers.[4] The member firms accounted for 5 percent of the mills in the United States and 1/3 of total production. The association collected and disseminated detailed price and quantity data from its members, inspected member firms, issued a report on changes in market conditions, and held monthly meetings of members. There may have also been recommendations not to increase production. Surveys of members indicated that some felt that the association acted to increase prices.

The Court found that the detailed disclosure of information was inconsistent with competition: "This is not the conduct of competitors, but is so clearly that of men united in an agreement, express or implied, to act together and pursue a common purpose under a common guide that, if it did not stand confessed a combination to restrict production and increase prices . . . that conclusion must inevitably have been inferred from the facts which were proved." In dissent, Justice Holmes and Justice Brandeis just as strongly argued that the sharing of information was consistent with competition.

Four years later the Court considered the same issues again. *Maple Flooring Manufacturers' Association v. United States* involved a twenty-two-member trade association that had 70 percent of the maple, beech, and birch flooring market. The association shared information about costs, freight rates, price, and quantity. It also had meetings of members to discuss industry conditions. The Court held that there was no violation of the antitrust laws.[5] Speaking quite generally, Justice Stone echoed the views of Brandeis and Holmes:

> We decide only that trade associations or combinations of persons or corporations which openly and fairly gather and disseminate information as to the cost of their product, the volume of production, the actual price which the product has brought in past transactions, stocks of merchandise on hand, approximate cost of transportation from the principal point of shipment to the points of consumption as did these defendants and who, as they did, meet and discuss such information and statistics without however reaching or attempting to reach any agreement or any concerted action with respect to prices or production or restraining competition, do not thereby engage in unlawful restraint of commerce.

In both these cases, the Court focused narrowly on the information that the associations gathered. This is not sufficient. Information about other firms in an industry can be put to both competitive and anticompetitive uses. Firms in a competitive market need to know what their rivals are doing. They must set their prices not by their own costs but by the prevailing market price. If one firm has developed an imaginative or more effective marketing strategy, consumers benefit when other firms are free to learn about it and imitate it. Information about rivals' prices, however, is also an essential element of successful collusion. To discover whether a trade association is promoting or hindering competition, one should focus not on information in the abstract, but on whether the structure of the particular industry is one in which tacit collusion is likely. If it is, one should look with suspicion on devices that make information about price readily available.

An important modern case on facilitating devices was brought by the FTC against the producers of ethyl chloride, a lead-based antiknock gasoline additive. The firms in the industry appeared to be very profitable despite the homogeneity of the product and a declining market. The FTC claimed that a number of industry practices facilitated cooperative pricing. The practices included public preannouncement of price changes, uniform delivered pricing, and most-favored nation clauses (if prices are decreased, recent purchasers must be rebated the difference).

These practices may in fact make tacit collusion more likely. Preannouncement of price changes allows rivals to react prior to any effect of price change on market shares. This reduction in reaction time implies that the appropriate period length is very short, so discount factors are high and collusion is easier; there is little ability to steal market share before competitors can react. Uniform delivered pricing, a practice in which all buyers pay a single price independent of transportation costs, can make it easier for firms to observe rivals' prices. It may therefore make it easier to respond to a price cut. Most-favored nation clauses reduce the incentive to cut prices because the price-cutter bears the additional cost of the rebates to previous customers. The reduction in gains from price cutting may make cooperative pricing easier to sustain.[6]

The court ruled that these practices were not violations of the antitrust laws because the existence of these practices was insufficient to show that an agreement had been reached among the firms. The implication is that, absent explicit agreements among competitors, facilitating practices are not illegal and the mere existence of these practices is insufficient evidence to prove agreement. As we have seen, however, an anticompetitive equilibrium can emerge even in the absence of explicit agreement. The justification for a legal rule in such cases then rests ultimately not on the absence of any ability on the part of firms to engage in tacit collusion, but rather on our inability to do much about it.

Price is merely one attribute among many. We might limit the ability of firms to change price, but we must also place other limits on these firms for the legal rule to be effective. In the airline industry, for example, one would have to confront the availability of different kinds of fares on each flight as well as the number of flights. One could limit the ability of airlines to respond to fare changes quickly by curtailing the use of computer reservation systems, but these systems have many benefits for consumers.

Parties who engage in tacit collusion are behaving quite differently from firms that enter into explicit cartels. The managers of firms that engage in tacit collusion may not even know what they are doing. They may, for example, use base-point pricing (basing shipping costs from a single source, even if it is not the source from which the goods will actually be shipped) because it is common practice; they may not recognize that the practice helps to support an anticompetitive equilibrium.

Reputation, Predation, and Cooperation

In a static environment, small firms can benefit when one of their competitors is a large monopolist with market power. The monopolist deliberately reduces output in order to keep price above the competitive level. The small firms may benefit because they too can charge the supracompetitive price. Things, however, are not always static. A frequent claim of small firms in an industry is that the large firm, instead of keeping prices high, will deliberately cut prices below cost for a time until the smaller firms are driven out of business, at which time it will return to charging the monopoly price. This practice is known as *predatory pricing*. For predatory pricing to be worthwhile, of course, the lower profits the predator earns during the predation stage must be more than made up by the increased profits after the prey has exited.

Many have argued that courts should be extremely reluctant to find that large firms have engaged in predatory conduct. What small firms see as predation may simply be the response of large firms to competition. Predatory pricing can work only if the implicit threat of the large firm (to continue low prices until the smaller firms leave) is credible. Because the larger firm sells more goods at the lower price than the small firm, predatory pricing is necessarily more costly to it. The threat to predate therefore may not be credible. A number of scholars have found only weak evidence for predation in those cases in which it has been alleged and have concluded that it may not make sense for the law to try to deal with it.[7] In their view, the benefits to consumers from confronting the problem of predatory pricing do not come close to matching the costs of litigating predation cases. Moreover, firms may charge high prices and leave consumers worse off merely because they risk being held liable for predation if they charge the competitive price.

To understand the legal debate, one must first understand how it might be rational for one firm to engage in predation. One of the best-known predation cases involved the practices of the American Tobacco Company.[8] The Supreme Court found that American Tobacco would enter a market of a much smaller firm, lower its prices dramatically—sometimes incurring substantial losses—persuade the smaller firm to merge with it, and then shut that firm down. By showing its willingness to incur such losses, it was argued, American Tobacco could persuade existing firms to sell out at artificially low prices and deter others from entering the tobacco industry in the first place. We want to ask whether one can explain how American Tobacco's threat to lower prices was credible, given that it was facing different small firms over time.

We can examine the problem of reputation-based predation by taking an entry and pricing game and repeating it over time. In each of the one-shot games, there are two players—Incumbent and Entrant. Initially, Entrant must decide whether to enter or stay out. If there is entry, Incumbent must decide whether to accommodate entry or predate. The payoffs are shown in the extensive form representation of Figure 5.4.

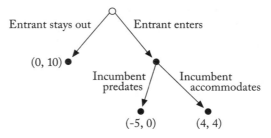

Figure 5.4 Predation and rational incumbent. *Payoffs:* Entrant, Incumbent.

We can solve the one-shot game through backwards induction. Once Entrant enters, Incumbent faces a payoff of $4 from accommodating and $0 from predating. Hence, Entrant believes at the start that Incumbent accommodates when the former enters. Because Entrant earns nothing by staying out and accommodating brings a payoff of $4, Entrant enters. When this game is repeated a finite number of times, there is a unique subgame perfect Nash equilibrium—Entrant enters and Incumbent accommodates. Incumbent would like to threaten predation to deter entry but the threat is not credible. In the last market, Incumbent accommodates in the event of entry because there are no other markets from which entrants might be deterred. Knowing that Incumbent will accommodate, Entrant decides to enter. In the second-to-last market, Incumbent accommodates in the event of entry because it knows that Entrant will enter in the next period regardless of what it does. Entrant, knowing this, will enter in the second-to-last market. This process of unraveling continues until the first market is reached. One can also imagine that Incumbent faces sequential potential entry in ten separate but identical markets where it is a monopolist. This repeated game, however, has the same outcome. The unique subgame perfect Nash equilibrium is entry and accommodation in each market.

If the game is of infinite or uncertain length and Incumbent faces the same entrant in every period, however, there are an infinite number of equilibria. If Incumbent faces a different entrant in every period, however, the only subgame perfect equilibrium is one in which there is accommodation in every period. Incumbent can punish any given entrant only after all that entrant's costs are sunk. Any given entrant is unmoved by Incumbent's threat to predate in future periods, and Incumbent always accommodates whenever it focuses exclusively on the one-shot game.

At this point, however, we return to the problem that we encountered in the installment sale game at the beginning of the chapter. The solution to the games of finite and indefinite length turns on the pay-offs' being common knowledge. The outcome may be altogether differ-ent if Entrant is only slightly unsure about Incumbent's payoffs from predating or accommodating. At this point, we want to show how we can build a model that captures this element of uncertainty.

Let us assume that there is a small chance that the payoff to Incum-bent from accommodating is not $4, but −$1. Incumbent, in other words, is one of two types. Most incumbents are "rational"[9] and they evaluate only the monetary returns from accommodating. A few in-cumbents, however, are "aggressive." They suffer a loss of face if they fail to carry out a threat and the profits they would earn from accom-modating are not enough to make up for it. We set out the one-shot game in Figure 5.5.

In this one-shot game, Incumbent predates when Entrant enters. In-cumbent prefers a payoff of $0 to one of −$1. For this reason, Entrant does not enter. Entrant prefers the $0 payoff from staying out to the −$5 payoff from entering. Incumbent's threat in this game is credible because Incumbent is moved by self-interest to carry it out when the occasion arises.

At this point, we want to consider the repeated game that arises when Entrant is not certain whether Incumbent is the rational type who receives the payoffs shown in Figure 5.4 or the aggressive type who receives the payoffs shown in Figure 5.5. Entrants in later markets ob-serve the previous actions of Incumbent. They can infer that if Incum-bent ever accommodated, Incumbent must be rational, but they cannot be certain when Incumbent predates whether Incumbent is aggressive or Incumbent is rational but is mimicking the actions of the aggressive type in order to deter entry. We illustrate the extensive form of the one-shot game in Figure 5.6.

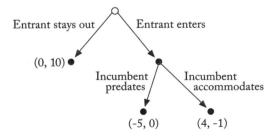

Figure 5.5 Predation and aggressive incumbent.
Payoffs: Entrant, Incumbent.

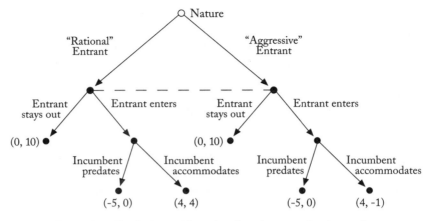

Figure 5.6 Predation with rational and aggressive incumbents. *Payoffs:* Entrant, Incumbent.

The equilibrium in the last period is easy to solve. The rational incumbent accommodates when Entrant enters and the aggressive incumbent predates. Entrant enters when it believes that the probability that Incumbent is rational is greater than 5/9 and stays out otherwise. The rational incumbent has no reason to predate. The rational incumbent, however, may choose to predate in early rounds in order to deter future entry by making potential entrants think that it may be an aggressive incumbent.

Entrant updates beliefs about Incumbent's type by looking at how Incumbent responded to entry in previous rounds. If Incumbent responded to entry in an early round by accommodating, future entrants will know with certainty that Incumbent is rational. From then until the end, backwards induction suggests that Incumbent will receive only $4 each period. If Incumbent responded to entry in early rounds by predating, however, future entrants will remain uncertain about whether Incumbent is rational.

If there are many periods left, it may be part of an equilibrium for the rational incumbent to predate. The equilibrium works in the following way: If Entrant expects both types to predate, Entrant will choose to stay out. We also need to know, however, about what happens off the equilibrium path. We must ask what happens if Entrant deviates and enters anyway. If Incumbent accommodates, future entrants know that Incumbent is rational, so they will enter. If Incumbent predates, future entrants are still in the dark. If it was worthwhile for both types of incumbents to predate in the preceding period, it may still pay for them to predate during the next period. If this is the case, Entrant will stay out. Therefore, by predating in the preceding period, the rational incumbent may deter future entry.

When there is some uncertainty in the payoffs, predation is a sensible strategy for both types of incumbents. The rational incumbent pools with the aggressive incumbent. The inability of entrants to distinguish the different incumbents deters entry. A key element of this argument is that a rational incumbent need not bear the costs of predation in the early periods. In equilibrium, there is no entry. If, however, there should be entry in the early periods, the rational incumbent will predate to maintain a reputation for aggressive behavior that deters future entry. One can extend this same idea to the infinitely repeated game.

This game is one way to model reputation. Incumbents who fail to predate reveal their type. They are willing to predate in order to convince future entrants that they are aggressive rather than rational. Indeed, predating is the way in which a reputation as an aggressive player is established. There are two important characteristics of equilibrium in the reputation game. First, the reputation equilibrium is unique, even if the game is infinitely repeated. We do not have the problem we encountered in other infinitely repeated games of multiple equilibria, only some of which lead to cooperative pricing. In this model, a reputation for aggressive responses to entry is the only equilibrium. Second, as we add periods, the fraction of periods in which entry is deterred approaches one, so the final periods where reputation does not work become economically unimportant.

The reputation-based model of predation suggests the possibility of successful, rational, predatory pricing. In addition to the reputation model, there are also signaling models of predation, in which an incumbent charges a low price to signal its low cost to an entrant or a potential entrant. The idea is to convince the entrant that competing with the incumbent will not be profitable, so the entrant should exit or not bother to enter. Another type of predation model is based upon the limited financial resources of the entrant rather than the entrant's cost disadvantage. If financial markets are imperfect, making it difficult for a firm to get outside financing, a protracted price war may force the financially weak firm out of business. This is sometimes known as the deep-pockets model of predatory pricing. The existence of three consistent models of rational predatory pricing—reputation, signaling, and deep pockets—does not mean that predatory pricing is rife. The contribution to antitrust policy from these models is to indicate what the necessary assumptions are to make predatory pricing work.

We have also developed the predation model in order to make a more general point. Models such as this one relax the assumption of common knowledge by introducing a little uncertainty into the payoffs. These show how patterns of behavior based on reputation might emerge in contexts in which folk theorems suggest that many equilibria exist. If only a few players have different payoff structures, the course of the repeated game for everyone may become radically different.

We have created a repeated game in which the problem was one of predation, but we could as easily have made the same point with a variation on the installment sale game in Figure 5.2. In that case, injecting uncertainty in the payoffs suggests how socially desirable cooperation might emerge. Even if only a few players value their reputations, it may be in the interest of everyone else to mimic them. Thus, in many situations, it may be possible to have cooperation for a long time as the unique outcome to the game even if it would be in the interest of most players to defect in the one-shot game. The problem is not explaining why most people would cooperate in the one-shot game, but rather why a few do.

Because the equilibrium may depend on the behavior of a few players who are unusual, rather than the players who are typical, it is worth spending a moment focusing on these players. The individuals who support these equilibria cannot simply be those who place a special value on the long-term. What matters is their payoff in the one-shot game. Their payoffs in the individual games must be such that, when they are called upon to carry out a threat or to cooperate, they will find it in their self-interest to do so. As we saw in Chapter 2, players may be able to make sunk investments that change the payoffs of the one-shot game. More generally, individuals may be able to change their payoffs in these repeated games by making investments now that will alter their own preferences in the future.

The preferences that an individual has will affect the payoffs that that individual receives in the one-shot game, and individuals can deliberately alter their preferences over time. Individual merchants, for example, might choose at the outset of their careers to enter a social circle whose members are those with whom they are likely to do business. Such a social circle might ostracize anyone thought to engage in sharp practices. A person might develop community ties and adopt a social life (such as by joining a church and other community groups) that would make it costly for that person to break a promise made to another merchant, even if that person were not subject to any legal damages.

To alter one's payoffs in the future, however, one must do more than feign attachment to a particular church or social group. One must actually form these attachments and suffer real costs from noncooperation. Some of these might be visible and some might not be. To the extent that these ties are verifiable, a particular individual can capture the entire benefit; to the extent that they are not, however, long-term relationships may be possible for all merchants. As long as some have genuinely committed to a particular social circle or a particular church group, others may find it in their interest to mimic them.

We can also imagine other ways in which players might be able to change the payoffs they receive in one-shot games. The owners of a firm, for example, might take advantage of the separation of ownership and control in many firms. They could hire managers who do not incur the costs of carrying out a threat. Various golden parachutes and other kinds of arrangements with managers may undercut their incentives to work hard, but they may also make managers more willing to carry out a threat. The profits of the firm would fall if the threat ever had to be carried out, but the compensation package may make the managers less concerned about profits and more concerned about preserving market share and gross revenues.

Once it is in the interest of a few managers to predate even when it is not in the interests of the owners, it becomes in the interest of other managers to mimic them (at least in early periods), even if their sole concern is profits. Here again, what matters is whether some firms are structured in such a way that the managers do not act in the interest of the firm when the time comes to predate and that these managers cannot be distinguished from others.

Legal rules can affect these cases of repeated interaction with small uncertainty in the payoffs in two different ways. First, legal rules can affect whether some people can make investments that change the way they play the one-shot game. For example, a legal rule that made it harder to exclude an individual from a social circle might have the unintended effect of reducing the informal sanctions for sharp practices and the uncertainty in the payoff from long-term cooperation. Without at least the possibility of such sanctions, long-term cooperation may become less likely. Second, legal rules may make it harder or easier for other players to mimic those whose self-interest leads to cooperation. Disclosure laws about the relationship between managers and firms may make it easier for players to identify the payoffs their competitors face and thus make it harder for ordinary players to be confused with those who enjoy a special set of payoffs.

Summary

Repeated games are inherently hard to analyze. In the first instance, we must be more cautious in making the assumption of common knowledge, and we must further ensure that the model reflects such things as the uncertainty in the length of the game and in the payoffs that the players receive.

When games are of uncertain length, folk theorems tell us that the mere possibility of repeated interaction does not ensure that cooperative behavior will emerge. Nevertheless, it takes only some uncertainty in the payoffs of the one-shot game for an equilibrium to emerge in the game that is in the interest of an individual player, and perhaps in the interest of society as a whole.

Legal rules designed to change the equilibrium of a repeated game may be ineffective. Players may have the ability to change the structure of the game. More subtly, legal rules may affect the equilibrium of a repeated game by changing the payoffs to a few players in the one-shot game. They may also make it harder for the mine-run of players to mimic those who have unusual payoffs.

Bibliographic Notes

Antitrust, neoclassical economics, and game theory. Both the traditional and the game-theoretic literatures on the economics of antitrust are extensive. One of the primary focuses of the field of industrial organization is to understand the effects and appropriate application of antitrust laws and other governmental regulation of competition. A large percentage of the applications of game theory has been in the field of industrial organization. Textbooks include Carlton and Perloff (1990), Scherer and Ross (1990), and Tirole (1988). Tirole's book emphasizes the game-theoretic aspects of modern industrial organization. Another useful reference is the collection of survey articles edited by Schmalensee and Willig (1989).

The limits of backwards induction. The idea that backwards induction leads to unraveling back to the first move is called the *chain-store para-*

dox because it was developed in the context of predatory pricing with an incumbent firm that owned a chain of stores in many towns and that faced a potential entrant in each town; see Selten (1978). The question faced was whether predation would be rational—backwards induction suggests that it would not be.

Infinitely repeated games, tacit collusion, and the folk theorem. An excellent discussion of the various folk theorem results is contained in Fudenberg and Tirole (1991a). The literature on the repeated prisoner's dilemma is quite large. A natural place to start is Axelrod's book (1984a). Stigler (1964) was the first to draw attention to the issue of detection and punishment as essential factors in the ability to support tacit collusion in a non-game-theoretic model. The idea was captured in a game-theoretic model by Green and Porter (1984). The model has been extended by Abreu, Pearce, and Stacchetti (1986). Also see Tirole (1988) for a simple and very accessible version of the model. Literature on the selection of equilibria in infinitely repeated games based on renegotiation includes Pearce (1988) and Farrell and Maskin (1989). Rotemberg and Saloner (1986) is an example of a paper that uses a selection criterion based on the most-collusive equilibrium to generate comparative static results on the ability of firms to collude.

Reputation and predation. The reputation model discussed in the text is based on Milgrom and Roberts (1982). The other pathbreaking reputation papers are Kreps and Wilson (1982) and Kreps, Milgrom, Roberts, and Wilson (1982). The facts of predation cases reported at the appellate level, however, do not provide evidence for this sort of predation. Nor is it easy to muster evidence that golden parachutes and the like are designed to give managers preferences that make the threat of predation credible; see Lott and Opler (1992). Ellickson (1991) discusses the interaction between legal and extralegal sanctions. For discussions of how individuals can alter their future preferences, see Becker (1993) and Sunstein (1993).

In addition to the textbook discussions, an excellent survey of models of predation is Ordover and Saloner (1989). The extent to which the theoretical possibility of predation translates into actual predatory behavior is a subject of controversy. See Easterbrook (1981) for a discussion of many of the issues.

CHAPTER 6

Collective Action, Embedded Games, and the Limits of Simple Models

The prisoner's dilemma and the stag hunt are often used to capture the problem of collective action. Individuals have the private incentive to take actions that are not in their joint interest. These two-by-two games are emblematic of the problem faced by a large group of individuals who cannot enter into binding agreements with one another. Since at least the publication of Mancur Olson's *The Logic of Collective Action* in 1965 and Garrett Hardin's *The Tragedy of the Commons* in 1968, the problem of collective action has been central to modern analyses of the role of the state. Legal rules are one way—but by no means the only way—of transforming these games to ensure that individuals have the incentive to take actions that are in the collective interest of society as a whole. These simple games therefore arguably provide sound theoretical underpinnings for legal intervention.

In this chapter, we return to the question of how individual and social interests can diverge and the role that law can play in aligning them. In the first part of the chapter, we briefly outline the basic issues at stake. A substantial literature on the problem of collective action has emerged, but, rather than survey or replicate it, we focus on the way legal rules can respond to the problems of collective action.

In the second part of the chapter, we show why care is required when we use game theory to understand these issues. It is one thing to identify a potential collective action problem and another to find the formal model that best captures it. A problem is often too quickly identified as a prisoner's dilemma. In some cases, the strategic interaction is in fact better modeled as a stag hunt or some other game. In many others, the modeler fails to understand the context in which a particular set of interactions takes place. Too often, the small, free-standing game is viewed as *the* game being played, when often the free-standing game is actually embedded in a much larger game. Embedding a small game in a large game often will have substantial implications for the play of the small game.

In the remainder of the chapter, we examine collective action problems that arise when information problems must be confronted as well. We draw on results from the mechanism design literature to show the limits of using legal rules to solve collective action problems in the face of private information. We then set out some recent results in private standard setting, the related issue of network externalities, and some results on information aggregation and herd behavior.

Collective Action and the Role of Law

Consider a prospective employee who is looking for a new job. The employee cares about the net wage paid, which, we will assume, is in turn determined by the cash wage paid, minus an amount representing the known risks associated with the job. Working with computers may lead to carpal tunnel syndrome, mining coal to black lung disease, throwing a baseball to a torn rotator cuff. Almost all jobs have some risks. In classical microeconomics, a competitive market would result in cash wages with full compensating differentials. If one job paid $15 per hour and had risks valued at $1 per hour, a second job that required the same skills with risks of $11 per hour—race car driving, say— should pay a cash wage of $25 per hour. Only known risks can be incorporated into these prices, however. Research in cognitive psychology suggests that low-level risks are commonly misestimated. Moreover, each employee is likely to invest too little in identifying and quantifying these risks.

From the standpoint of the employees as a group, information about the workplace may be a public good. Once acquired, the information is something that is available to all at little or no cost. Because the producer of the information may have no control over it, no one may have the incentive to gather it in the first place. Instead of gathering the information, each employee would rather benefit from other employees' investments in information. This is the classic free-riding problem, which leads to too little effort being invested in learning about the company.

We should not assume, however, that legal intervention is necessary to ensure that the employees are given the information they need. The problem, after all, is one of disclosure. As we saw in Chapter 3, when information is verifiable (as it must be if it is to be the subject of direct government regulation), we have to ask whether there will be unraveling. We need to ask whether the safest employers reveal that they are safe, those that are only a little less safe follow suit, and so forth. Even if firms could not themselves credibly disclose the information, cheap communication devices could emerge which would solve the free-riding problem. Underwriters' Laboratories, probably the best-known private certifier of quality, tested more than 74,000 products in 1990. The UL label on a product is a simple statement that the product conforms to UL's exacting standards.

In similar fashion, employers could contract with private firms to certify how safe particular workplaces are, just as firms certify their financial health using outside auditors, or product manufacturers certify safety through the UL label. Employers might seek such a certification so as to be pooled as a group and to differentiate themselves from uncertified employers. Employers would finance the system, just as product manufacturers currently do for other forms of labeling. The willingness of a firm to submit voluntarily to an outside audit is a standard device for signaling the firm's private information that, from the point of view of outsiders, is nonverifiable. And the labeling scheme greatly simplifies the prospective employee's decisionmaking problem. The label is there or it is not, and the employees no longer face the free-riding problem that would otherwise exist and that would lead to too little investigation into information.

Legal rules may be needed to address such collective action problems when, for some reason, these private devices do not work. There seem to be two different kinds of approaches. First, the government itself can overcome the collective action problem by gathering the information about the appropriate safety standards and ensuring that individual employers meet them. The government stands in the shoes of all the employees as a group, both gathering information about safety and ensuring that employers comply with its regulations. This rationale might justify OSHA and consumer product safety legislation.[1]

Approaches that directly regulate activities, however, suffer from a familiar disadvantage. They may make unreasonable informational demands on the court, regulator, or other person enforcing the law. The more information the person needs to know to design a rule or enforce it, the less confident we can be that it will work effectively. The second approach available to a lawmaker is to expand or otherwise alter the strategy space of the players in such a way that information that would otherwise not be disclosed is disclosed. The National Labor Relations

Act, for example, allows employees to organize themselves and in that way internalize the costs of gathering information. The differences between these two approaches are analogous to those that we discussed in Chapter 4 when we examined the differences between laws that affect the actions of the informed player who sends a signal and those that affect the actions of the uninformed player who receives it.

Embedded Games

Many problems involving simultaneous decisionmaking are actually embedded in larger decisionmaking problems. Hence, before we can be sure that a simple game captures the dynamics of a collective action problem or any other complicated interaction, we must understand the extent to which it can be isolated from the context in which it arises. To get at these ideas in a straightforward setting, we will consider a series of stylized examples of games and embedded games. Although these models abstract significantly from actual situations, the simple settings highlight the importance of embedding. In particular, the examples should make clear both the substantial risk of misidentification of the game and the possible constructive role of legal rules in defining game structure. Return to the basic two-by-two normal form game structure we saw in Chapter 1. Two players make simultaneous, independent decisions. The payoffs to each depend on the decisions of both. Many of the best-known noncooperative games are particular cases of this general form. Everything turns, of course, on the relationships among the payoffs.

Figure 6.1 illustrates a prisoner's dilemma and a coordination game. Each player must decide which strategy to adopt independently of the other. As we saw in Chapter 1, in the prisoner's dilemma each player

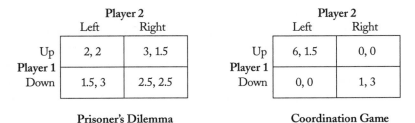

Figure 6.1 Simple collective action games. *Payoffs:* Player 1, Player 2.

has a dominant strategy, and therefore the play of the game is readily predictable. Player 1 adopts the strategy of up, and Player 2 adopts the strategy of left. Each enjoys a payoff of $2.

The coordination game has two pure strategy Nash equilibria, the strategy combinations of up and left and down and right respectively. The strategy combination of up and left forms a Nash equilibrium. Player 1 cannot do better than to play up when Player 2 plays left, and Player 2 cannot do any better than to play left when Player 1 plays up. A similar argument holds for the strategy combination of down and right. Because of these multiple equilibria, we cannot predict the outcome of the coordination game.

The situation changes, however, if the coordination game is embedded in a larger game. Let us assume that, in this larger game, Player 1 makes an initial move in which Player 1 has a chance to decide between taking a certain payoff of $2 or playing a coordination game. We illustrate this game in two different ways in Figure 6.2. The second illustration is the same as the first, except that it replaces the normal form subgame with the corresponding extensive form game. Player 1 can play left and end the game with payoffs of $2 to each player or Player 1 can play right. In that event, both players play the coordination game we set out in Figure 6.1.

Player 2 decides whether to play left or right only after observing that Player 1 has moved right. Player 2 does not know whether Player 1 moved up or down, but Player 2 should not expect Player 1 ever to move down after having moved right. Player 1 receives a maximum payoff of $1 by playing right and then down. Player 1 always receives $2 by moving left. As long as Player 2 thinks that Player 1 will never play a dominated strategy, Player 2 must believe that Player 1 never chooses the strategy of right followed by down.

Anytime Player 2 sees that Player 1 moves right, Player 2 should believe that Player 1 will then move up. For this reason, Player 2 plays left in response to Player 1's initial move of right. Because Player 2 believes that Player 1 will move up after moving right, Player 2 ensures a payoff of $1.50 rather than $0 by moving left. Player 1 has two beliefs: First, that Player 2 is rational and, second, that Player 2 believes that Player 1 is rational and therefore does not play dominated strategies. Hence, Player 1 infers that Player 2 moves left. For this reason, Player 1 adopts the strategy of first moving right and then up. By doing this, Player 1 enjoys a payoff of $6 instead of the payoff of $2 that Player 1 would receive from moving left initially. Even though this coordina-

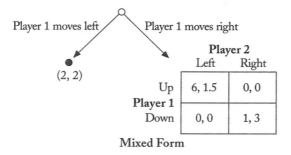

Mixed Form

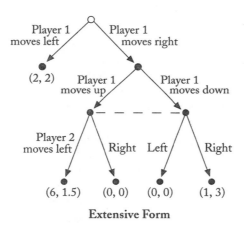

Extensive Form

Figure 6.2 Embedded coordination game. *Payoffs:* Player 1, Player 2.

tion game standing alone does not have a unique solution, it does have one when it is part of a larger game.

Coordination can also arise in more complicated settings. The game shown in Figure 6.3 embeds both a prisoner's dilemma and the coordination game. Player 1 moves left or right. After left, Player 1 and Player 2 confront a prisoner's dilemma; after right, they confront a coordination game. As before, Player 2 knows whether Player 1 chose left or right before moving. This game should have the same solution as the game in Figure 6.2. Indeed, the only difference between the two is that a prisoner's dilemma (the solution of which leaves each player with a payoff of $2) replaces a terminal node in which each player again receives a payoff of $2.

Once again, Player 1 would never move right and then play down.

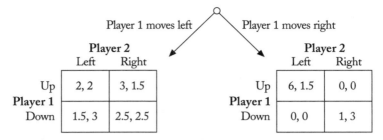

Figure 6.3 Embedded prisoner's dilemma and coordination games. *Payoffs:* Player 1, Player 2.

Such a strategy leads to a payoff of $1 or $0. Player 1 should not play this strategy given that another strategy (moving left and then up) has a certain payoff of $2. Player 2 should therefore infer that, if Player 1 ever chooses to play right, Player 1 will follow this move by playing up. Hence, Player 2 adopts the strategy of playing left. Because Player 2 chooses this strategy, Player 1 chooses right.

If either of the subgames that begin after Player 1 makes the initial move were free-standing, there would be either an inefficient outcome (in the case of a prisoner's dilemma) or an indeterminate one (in the case of the coordination game). When these games are part of a larger game, however, the outcome may be altogether different. In Figure 6.3, the existence of a scenario in which a prisoner's dilemma game might arise actually helps the players to achieve the outcome that is in their joint interest. Game structure matters, and often matters a lot. Identification of the game itself is of great importance. Misidentification usually occurs when the small, free-standing game is viewed as *the* game. A modeler who focused on the interaction captured in the prisoner's dilemma in Figure 6.3 rather than on the entire game would be misled.

This problem of embedded games always confronts someone concerned with the problem of analyzing legal rules. Laws never work in isolation, and too much may be lost when we try to focus on one aspect of the problem. The Trust Indenture Act, for example, makes it difficult for holders of publicly issued debt to renegotiate the terms of their bonds when the debtor is in financial distress.[2] Because of changed conditions, the debtor has a suboptimal capital structure, but no individual bondholder will restructure the debtor's debt. As a result, the bondholders lack any way of working together collectively to alter the terms of the debt. The bondholders face a collective action problem.

Viewed in isolation, this feature of the law might seem a weakness. Indeed, one can justify the bankruptcy laws and particularly the willingness of bankruptcy judges to confirm prepackaged plans on the ground that they solve this collective action problem. One can argue, however, that the existence of this collective action problem is a good thing. It alters the shape of the other negotiations that the debtor faces and changes the incentives of the debtor and others during the course of the relationship. These effects, of course, could be harmful as well as beneficial, but they need to be taken into account before one can be confident that eliminating this collective action problem is desirable.

This observation can be extended to many other legal contexts. Civil litigation itself is at best a zero-sum game for the parties. When viewed in isolation, it seems inherently wasteful because all the resources that are being spent merely alter relative payoffs to the plaintiffs and the defendants. Civil litigation, however, is by definition an embedded game. One cannot understand the role that it plays without placing it in its larger context. Reducing the costs of litigation affects not only the probability of settlement but also the way parties act in the first instance. Sorting out the relationship between the litigation game and the other aspects of the interactions between the parties is especially difficult in cases such as class actions, in which there may be multiple plaintiffs or multiple defendants.

Every contractual relationship poses similar problems. When one or another party defaults, the situation in which both parties find themselves may take the form of a prisoner's dilemma, but it might be that the existence of this prisoner's dilemma prevents the default from happening in the first place and ensures that the parties take actions that are in their self-interest.

Understanding the Structure of Large Games

In the previous section, we saw how we might not understand the likely course of play if we examined interactions in isolation. An isolated prisoner's dilemma is a problem; an embedded prisoner's dilemma may solve a problem. Not seeing the larger game leads to a mistaken analysis. Yet missing part of the game is only one of many mistakes that can be made. As we show in this section, it is easy to misidentify the game being played. What appears to be a prisoner's dilemma may be something else entirely. We must be cautious about concluding that the insights of a particular game-theoretic model can be carried over to a particular real-world situation, especially in cases in which the parties have a contractual relationship with one another. For example, we should hesitate before concluding that lawmakers should confront the collective action problem that bondholders seem to face when confronted with an exchange offer or the problem that shareholders face in the wake of a tender offer, given their ability to bargain before the fact. This line of inquiry also forces us to question the justifications that are commonly offered in support of bankruptcy laws.

A common approach to bankruptcy scholarship offers a good illustration of the need to exercise caution before characterizing any particular interaction as a paradigmatic game such as a prisoner's dilemma. Much of the bankruptcy scholarship over the last decade is premised on the idea that bankruptcy law exists to solve a common pool problem. Individual creditors pounce on the failing debtor. They ignore the larger consequences of their actions when their debtor is in financial straits. Left to their own devices, individual creditors would tend to seize assets in an effort to be paid in full; they would not care whether the assets would be taken from their best use and might be careless in disposing of them. Compounding this problem, individual creditors would spend time and money ensuring that others did not act before them. Because creditors face high costs in negotiating with one another, the explanation goes, they face a common pool problem. We need bankruptcy law, under this view, because it imposes a mandatory collective proceeding to prevent individual creditors from acting in a way that is contrary to the interests of creditors as a group.

This problem, however, is different from most other common pool problems that the law addresses. These other problems are typically the domain of property law. The common pool problem arises among strangers who have no relationship with each other or with any common third party. By contrast, the relationships among the debtor and its creditors are largely contractual. The parties themselves can structure their relationships with one another to minimize the common pool problem. Firms often have both secured and unsecured creditors; still others grant no secured debt but have senior and subordinated unsecured creditors. By the initial allocation of priority rights, many firms can avoid the common pool problem altogether. In other words, the game among the creditors that results when the firm fails is embedded in the much larger decisionmaking game about the firm's capital structure.

Debtor has a chance to undertake a new project. The expected value of the project is $115. (In 20 percent of the cases, the project yields $84;

the remaining 80 percent of the time it yields $122.75.) Debtor, however, must borrow $100 in order to have enough capital to undertake the project. To borrow that much money, Debtor must obtain loans from 2 creditors.[3] The lending market is competitive and has a risk-free rate of return of 0 percent.

We consider first a legal regime in which security interests do not exist. The only decision the creditors must make is whether to monitor Debtor after lending the money. A creditor who monitors can detect whether the project has failed. Monitoring after the loan is made costs a fixed amount. Creditor 1's cost is $5 and Creditor 2's cost is $8. If only one creditor monitors Debtor and the project fails, that creditor can successfully withdraw its money from the firm and receive payment in full. If both creditors monitor and the firm fails, the assets of the firm are distributed pro rata. Creditors also share pro rata if neither monitors Debtor.

In this model, monitoring brings no social benefits. The outcome that is in the joint interest of the creditors is one in which neither monitors Debtor. For this reason, the creditors and the debtor at the outset should find it in their interest to create a set of obligations such that neither creditor would ever be better off monitoring, regardless of what either creditor thought the other would do. We should therefore start by asking whether the parties can create a set of payment obligations such that neither creditor monitors and both expect to be repaid. We do this by asking whether the combination of strategies in which neither monitors Debtor is an equilibrium.

If a combination of strategies in which neither creditor monitors the debtor is an equilibrium, the subgame that begins after each lender extends credit takes the form shown in Figure 6.4. In this proposed equilibrium, neither creditor monitors; hence, the expected return to each creditor when neither monitors must be $50, given our assump-

	Creditor 2	
	Don't Monitor	Monitor
Don't Monitor **Creditor 1**	50, 50	48, 44
Monitor	47, 48	45, 42

Figure 6.4 Creditor common pool (low-risk debtor).
Payoffs: Creditor 1, Creditor 2.

tion that the competitive rate of return is 0 percent. If neither creditor monitors, each receives only $42 when the project fails. The project fails 20 percent of the time. To enjoy an expected return of $50 and make up for the $8 loss they suffer when the project fails, the creditors must be paid an extra $2 (or $52) in the 80 percent of the cases in which the project succeeds.

In this proposed equilibrium, Creditor 1 receives $50 when it does not monitor. Given that Creditor 2 is not monitoring, this is a best response. If Creditor 1 monitored, it would receive $52 regardless of whether the project succeeded or failed, but it would have spent $5 on monitoring. Its net payoff would be only $47. When Creditor 1 does not monitor, not monitoring is also a best response for Creditor 2. Monitoring would again guarantee Creditor 2 a return of $52, but it would have to spend $8 on monitoring for a net return of only $44. Because the strategy of each is a best response to the strategy of the other, the strategy combination in which neither monitors is a Nash equilibrium.[4]

Consider the case that would arise if the project still had an expected value of $115 but only a 50 percent chance of success. The project yields $146 if it is successful and $84 if it fails, giving again an expected value of $115. We can test whether an equilibrium exists in which neither creditor monitors. If such an equilibrium exists, the creditors again would set their expected returns to equal $50 when neither monitors. To do this, each creditor must receive an $8 premium (or $58 in all) when the project succeeds to offset the $8 loss in the equally likely event that the project fails. In Figure 6.5 we show the subgame that would begin if both creditors lent on the belief that neither would monitor.

We can now test whether the proposed strategy combination in which neither monitors is an equilibrium. Given that Creditor 2 is not monitoring, Creditor 1 receives $50 when it does not monitor. Creditor

| | | Creditor 2 | |
		Don't Monitor	Monitor
Creditor 1	Don't Monitor	50, 50	42, 50
	Monitor	53, 42	45, 42

Figure 6.5 Creditor common pool (high-risk debtor; proposed equilibrium with no monitoring). *Payoffs:* Creditor 1, Creditor 2.

1, however, receives $58 if it monitors—regardless of whether the project fails or succeeds—less its costs of monitoring. Monitoring costs Creditor 1 only $5. Hence, Creditor 1 receives a net payoff of $53 when it monitors. Because this payoff is greater than the $50 payoff from not monitoring, the strategy combination in which neither player monitors is not an equilibrium.

Another proposed equilibrium to consider is one in which both creditors monitor. In this proposed equilibrium, the fees are set at the beginning so that both creditors have an expected payoff of $50 when each monitors. Because both creditors now incur the costs of monitoring in equilibrium, the debtor must pay them enough when the project succeeds to compensate them not only for the risk of failure but also for the costs of monitoring. Creditor 1 must be paid $69 when the project succeeds. It must be compensated both for the possibility that it will not be paid in full and for the costs of monitoring. Creditor 2 must be paid $73 because its costs of monitoring are even higher.[5] We show the subgame that would arise after each extended credit in this proposed equilibrium in Figure 6.6.

Creditor 1 receives $50 in this proposed equilibrium. Creditor 1 could save $5 in monitoring costs if it did not monitor. In this event, it would receive $69 when the project succeeds, which happens ½ the time, but it would receive only $11 when the project fails, which again happens ½ the time. (The $11 is the amount left from the $84 after Creditor 2 grabs $73.) Creditor 1's expected payoff from not monitoring is therefore $40. Its best response is thus to monitor when Creditor 2 monitors.

Creditor 2 is in a similar situation. It receives $73 when the project succeeds, but only $15 when it fails. (This is the amount that is left of $84 after Creditor 1 takes the $69 it is owed.) These happen with equal probability, and therefore Creditor 2's expected payoff from not moni-

	Creditor 2	
	Don't Monitor	Monitor
Don't Monitor	55, 58	40, 65
Creditor 1		
Monitor	64, 44	50, 50

Figure 6.6 Creditor common pool (high-risk debtor; proposed equilibrium with monitoring). *Payoffs:* Creditor 1, Creditor 2.

toring, given that Creditor 1 monitors, is $44. When Creditor 1 monitors, monitoring is a best response for Creditor 2 because it brings a payoff of $50 rather than $44. Because monitoring is a best response to the strategy of the other player, this combination is a Nash equilibrium.

In this world in which there is only unsecured credit, there is a common pool problem. Low-risk debtors will not be monitored at all—a no-monitoring equilibrium—because the cost of monitoring exceeds the benefits gained from it. (The benefits of monitoring are either grabbing more than a pro rata share of the assets or ensuring that one receives such a share.) Both creditors monitor high-risk debtors. Because monitoring is costly, interest rates are in turn higher because the lenders must be compensated for the costs they incur in making the loans. In the end, the forces that lead the creditors to monitor lower the value of the debtor's equity. In this model, monitoring creates no efficiencies and simply allocates value between the creditors.

Introducing the possibility of secured credit changes the analysis. Issuing secured credit to one of the creditors may eliminate the incentives for either creditor to monitor. We can see this by testing whether it is an equilibrium for neither creditor to monitor when Creditor 1 has a right to be paid before Creditor 2. In Figure 6.7 we show the subgame that would arise in such an equilibrium after credit is extended.

When Creditor 1 takes a security interest and Creditor 2 does not, neither creditor gains anything from monitoring. The strategy combination in which neither monitors is a Nash equilibrium. If either creditor monitored, that creditor would receive the same amount of money from the debtor, but would be worse off because it has incurred the costs of monitoring and these costs must be subtracted from the payoffs. A creditor is always worse off monitoring. Therefore, not monitoring is always a best response when a creditor takes security. The strat-

| | Creditor 2 | |
	Don't Monitor	Monitor
Don't Monitor	50, 50	50, 42
Monitor	45, 50	45, 42

Figure 6.7 Creditor 1 takes security interest (high-risk debtor; no monitoring). *Payoffs:* Creditor 1, Creditor 2.

egy combination in which neither monitors but one takes security is a Nash equilibrium. The high-risk debtor prefers to grant one creditor security. By granting security, the debtor is able to obtain financing without having to pay Creditor 1 an additional $5 and Creditor 2 an additional $8 to cover their costs of monitoring, the outcome that we would see if the debtor were not able to give one creditor priority over another.

A legal regime that allows debtors to grant security or collateral may therefore make everyone jointly better off if the cost of obtaining a security interest is less than the total monitoring costs of the two creditors. In the case of a low-risk debtor, neither creditor will monitor. In the absence of the risk of such monitoring, neither creditor will take a security interest. One creditor, however, will take security when the debtor is high-risk in order to ensure that neither has the incentive to monitor the debtor's insolvency. In short, this model predicts a pattern of secured credit in which low-risk firms do not use secured credit, while high-risk firms do.

The priority that a security interest gives one creditor minimizes the opportunities for end-of-game efforts to grab assets. When a secured creditor is in the picture, there are no assets for general creditors to grab. Hence, the existence of secured credit may itself eliminate the common pool problem. This model suggests that the common pool problem is not an immutable feature of the relationship between a debtor and its multiple creditors. The prisoner's dilemma game thought to exist among unsecured creditors of the failing firm may be largely avoided through the initial contracts made in creating the capital structure of the debtor.

In the previous section, we showed how the presence of a prisoner's dilemma inside a larger game does not in itself necessarily generate an inefficient outcome of the game as a whole. In this section, we showed that an interaction that appears to be a prisoner's dilemma or some other paradigmatic game may in fact be part of a larger game that has a completely different structure. Bankruptcy law may not have to contend with a common pool problem at all. The most important effect of the Bankruptcy Code may be the way that it sets a framework for negotiations between the debtor and its principal creditor when it becomes insolvent. We explore the dynamics of these negotiations in the next chapter. The ability of parties to solve the collective action problem contractually may do much to explain why other legal systems appear to have responded to the problem of financial distress so differently than the United States. If the problem that needs to be addressed is not a collective action problem, we should expect the common characteristics of the different regimes to lie elsewhere.

Collective Action and Private Information

We saw in Chapter 1 that, when information is complete, it is possible to ensure that parties have incentives to act in a way that is in their joint interest. A law can provide for a schedule of damages for each possible combination of strategies that ensures that the socially optimal combination is the one that the parties choose. In virtually all cases that we encounter, however, we must contend with private information. For this reason, we must ask how laws can be crafted to take account of the existence of private information. In this section, we examine the special difficulties that arise with collective action problems when there is private, nonverifiable information.

The government of a town with a population of 100 has to decide whether to spend $104.50 to improve the landscaping in the municipal park. The new trees and flowers are a public good. The park is open and it would be prohibitively costly to charge a user fee to those who walk through it or drive by it. Some residents value the new landscaping more than others, but how much a particular individual values it is private, nonverifiable information. The town officials want to spend the money on the landscaping only if the residents as a group value it more than $104.50. These officials would like to devise some mechanism to find out whether the landscaping is in fact worth this much.

Assume, for simplicity, that each resident values the landscaping at $2, $1, or $0. Each resident's value is independent of the other residents' value. The simplest approach is to put the question to a vote and undertake the project if more than a certain fraction of the residents vote in favor of it. If the project is approved, each resident will be assessed a tax of $1.045 (the total cost of the improvement divided by the number of residents).

Putting the matter to a vote, however, does not provide the government with sufficient information to make the efficient decision. Only those residents who value the improvement at $2 will vote in favor of it; those who value the improvement at $1 or $0 prefer no landscaping, given the $1.045 tax they must pay if the landscaping is approved. By putting the matter to a vote, the town officials learn what fraction of the population values the project at $2. To make the correct decision, however, the officials also need to know what fraction values the project at $1. For example, if 30 people vote in favor of the improvements, we can infer that 30 people value the project at $2 and that the landscaping is worth at least $60. If 45 or more of the 70 people who voted "no" value the improvement at $1, however, the project is in the joint interest of the community and the vote tells us nothing about this.

Another possibility is to ask the residents to state the value that they place on the improvement. Each resident fills in a ballot and indicates a valuation of $0, $1, or $2. If the total value exceeds $104.50, then the improvement is implemented. If the improvement is made, each resident is assessed a tax of $1.045. This approach to the problem does no better than the simple vote.

Those who value the improvements at $2 will report a value of $2. If the landscaping is done, they will receive a $2 benefit and will have to pay a tax of only $1.045. Those who place a value on the improvements of $0 will report a value of $0. They do not want to pay anything for something they do not value. Consider, however, a resident's voting strategy if the resident values the landscaping at $1. If the landscaping is done, the resident will be assessed a tax of $1.045 and will receive a benefit of only $1. For this reason, the resident is worse off reporting a valuation of $1 or $2 rather than a value of $0. Therefore, those who value the landscaping at $1 report a value of $0. The only information that the town officials learn is the same as with a simple vote—the number of residents who value the landscaping at $2.

Both of these schemes fail because the taxes that are paid are not tied to the value that the residents receive from the landscaping. As long as this is the case, some residents will not have the incentive to report their true values. Those who value the project more than its per capita cost have an incentive to overstate their valuations, and those who value the project less than its per capita cost have an incentive to understate their valuations.

It is possible, however, to have schemes that do lead individuals to state their true valuations. The difficulty that such schemes must overcome is the familiar free-rider problem. For example, if we vary the amount of the tax according to the valuation that each resident states, each resident has an incentive to understate the value that resident receives from the landscaping. Stating the valuation truthfully when it is high makes sense only when telling the truth affects whether the landscaping is done. If all the other valuations exceed $104.50, a resident who states a positive valuation incurs a tax by doing so and receives no benefit, as the landscaping will be done anyway. To avoid the free-rider problem, we must ensure that a resident's tax liability increases as a result of a truthful valuation only in situations in which the truthful valuation changes the outcome. In this way, stating a truthful valuation brings a benefit to each resident that more than offsets the increased tax liability that resident faces.

One such mechanism that gives all the residents the right incentives works in the following way.[6] Residents again indicate whether they place a value of $0, $1, or $2 on the landscaping. The landscaping is done when the total valuation is greater than $104.50. Taxes, however, will be assessed against each resident by looking first at the sum of the valuations of all the other players. If this sum is less than $104.50, but close enough to it that it exceeds $104.50 when we add this resident's reported value to it, we impose a tax on the resident. This tax is equal to $104.50 less the sum of the reported values of all the other players. In all other cases, a resident pays nothing.

This scheme is built on the idea that a resident wants the landscaping done whenever the benefit that resident derives from the project is greater than the tax the resident has to pay. A resident's reported value affects whether a tax is paid, but once the reported value is high enough to trigger liability, the amount of tax that is paid is independent of the reported value. Under this tax assessment scheme, the weakly dominant strategy for each citizen is to report the true value for the landscaping.

We can see why each resident has the incentive to report the true value by looking at each possibility in turn. Consider situations in which the reported value does not affect whether the landscaping is done. This will occur when the sum of the reported values for the other players either exceeds the project cost or is so much less than it that no value the resident could report would make a difference. In either event, the resident pays no tax. Whether the landscaping is done has nothing to do with the value that the resident reports. In such a situation, reporting the true value is as good as (but no better than) any other reported value.

We need, then, to focus on the possibility that the sum of the reported values of the other players falls just short of the cost of the landscaping. The resident's reported value may determine whether the landscaping is done and may trigger a tax. The tax is the difference between the project cost and the sum of the other reported values. There are two possibilities. First, the tax may be less than the benefit that the resident derives from the landscaping. If it is, the resident is better off stating a value that is large enough to ensure that the park is built. Any value greater than the difference between the cost of the landscaping and the sum of the valuations of the other residents ensures that the landscaping is done. The true value that the resident attaches to the landscaping is the only value that always does this, regardless of the values of the other residents. Truthfully reporting the value is better than reporting a value that is so low that the landscaping is not done, and it is no worse than reporting any other valuation.

Second, the tax may exceed the benefit the resident receives from having the park built. This would occur if the sum of the valuations of the other residents falls far enough short of the cost of the landscaping that the difference between this sum and the cost of the landscaping is greater than the benefit that the resident would receive from the landscaping. In this case, the resident reports a value that is low enough to ensure that the landscaping is not done. The resident could do this by stating a value of $0; stating the true value, however, works equally well. (The true value is the benefit that the resident derives from the landscaping, and, in the case we are examining now, this amount is less than the difference between the cost of the landscaping and the sum of the reported values of the other residents.)

In short, regardless of the values that the other residents state, a resident is never any worse off stating the true value and may, in fact, be better off. Stating the true value is a weakly dominant strategy under this scheme.[7] This scheme works because it gives each individual the right marginal incentive in situations where that resident's reported value matters and provides no reason for the individual to lie in situations where the resident's reported value does not. This is known as a *Clark-Groves mechanism*. Note that individuals do not need to know the distribution of valuations to make this work because they have dominant strategies. Nor do the town officials need to know the distribution of valuations. The only information they need to know is the cost of the project.

These mechanisms, however, have a notable defect. Government receipts and outlays may differ widely. Assume, for example, that the total valuation of all the residents is $110. Because no one's vote is pivotal, no one is assessed a tax. The government has a deficit of $104.50. To fund the landscaping, the government must finance the project out of general revenues. These mechanisms stand in strong contrast to regimes of civil damages, in which individuals have the incentive to engage in actions that are in the joint interest of everyone and there is always a balanced budget.

In Chapter 1, we showed that the socially optimal outcome could always be obtained with dominant strategies through a regime in which individuals were given the right to sue others for damages. This result, however, depended crucially on our assumption of complete information. There is a general lesson of the literature on Clark-Groves mechanisms as applied to legal analysis: When information is incomplete, regimes in which individuals can bring damages actions against others are generally not sufficient to bring about socially optimal outcomes through dominant strategies.

It is possible to create a mechanism that leads the residents in our example to state the value they place on the park and at the same time maintain a balanced budget. Under such a mechanism, however, the players no longer have dominant strategies. Instead, the equilibrium is a Bayesian Nash equilibrium. This mechanism depends crucially on the players' knowing the distribution of valuations in the population. Because it makes stronger assumptions about what information is available than a Clark-Groves mechanism, it is harder to implement.

To see how such a mechanism might work, we shall simplify the example. There are only two residents and the cost of the improvement is $2.70. Again, the valuations can be $0, $1, or $2. The players believe that the value each attaches to the landscaping is independent of the other's valuation. In addition, each player believes that a player is as likely to have one valuation as another. The chance that a player has any particular valuation is one-third. We can implement the socially desirable outcome with the following scheme. If both players state a valuation of $2, they share the cost of the landscaping. If they both state $0 or both state $1, or if one states $0 and the other $1, they pay nothing and the landscaping is not done. If one player states $1 and the other states $2, the landscaping is done and the player who states $1 pays $1.25 and the player who states $2 pays $1.45. If one player states $2 and the other states $0, then the landscaping is not done and the player who states $0 pays the other player $0.30.

Under this scheme, an equilibrium exists in which both players tell the truth. Assume that Player 2 reports truthfully. Given the initial distribution of values, this means that Player 2 reports $0, $1, and $2 with equal probability. What is Player 1's best response to this? If Player 1 reports a value of $0, the landscaping is never done. Player 1 receives no benefit and pays no tax. Because Player 2 reports a valuation of $2 $\frac{1}{3}$ of the time, Player 1 must pay Player 2 $0.30 one third of the time. The rest of the time, Player 1 pays nothing. Hence, reporting a value of $0 has an expected value to Player 1 of −$0.10.

Reporting a value of $1 gives Player 1 nothing in $\frac{2}{3}$ of the cases. (Player 1 receives nothing when Player 2 reports $0 or $1. The total of both reported values in such a case is less than the project cost of $2.70.) In the remaining ⅓ of the cases in which Player 1 reports a value of $1, the project is undertaken. The expected value (net of taxes) from reporting $1 is therefore −$0.42 when Player 1 has a true value of $0, −$0.08 for a true value of $1, and $0.25 for a true value of $2. Finally, if Player 1 reports $2, ⅓ of the time Player 1 receives a side-payment of $0.30, and ⅔ of the time the project is built. Given the tax and side-payment schemes, the expected value from reporting $2 is −$0.83 for a true value of $0, −$0.17 for a true value of $1, and $0.50 for a true value of $2. This is summarized in Figure 6.8.

Player 1's best response to truthful reporting is truthful reporting. If Player 1 has a value of $0, the best that Player 1 can do is receive an expected payoff of −$.10. Player 1 receives this by truthfully reporting a value of $0. If Player 1 has a true value of $1 and the other players report truthfully, Player 1's best expected payoff is −$.08. Player 1 enjoys this expected payoff by reporting a value of $1. If Player 1 has a true value of $2, Player 1's best expected payoff is $0.50. Player 1 enjoys this payoff by stating a value of $2. Because the game is symmetric, Player 2 also has truth-telling as a best response when Player 1 tells the truth.

The need to use the Bayesian Nash solution concept, however, runs counter to the premise on which we began. This solution concept depends upon the players' sharing a common set of beliefs about the distribution of various types in the population. If this distribution of beliefs is known to the players, it may be known to the town officials as well. If the town officials know the distribution of beliefs, they should be able to determine whether the residents as a group believe that the benefits of the landscaping are worth its costs, provided that

	True Value		
	$0	$1	$2
Report $0	-$0.10	-$0.10	-$0.10
Report $1	-$0.42	-$0.08	$0.25
Report $2	-$0.83	-$0.17	$0.50

Figure 6.8 Payoffs from reporting different values to a player with a true value of $0, $1, or $2 when the other player reports truthfully.

the population is sufficiently large and each individual's valuation of the landscaping is independent of every other's. (In fact, weaker assumptions will often suffice.) This solution concept does not depend upon any player's knowing the valuation of another, but it does make strong assumptions about the information that the players possess.

These schemes suffer from a further problem: they cannot be implemented voluntarily. In our example, a player with a valuation of $0 or $1 suffers an expected loss from playing this game. The expected benefit to the player with a $2 valuation more than makes up for the loss, but the players with the $0 or $1 valuations know who they are before the game begins. Hence, they will not play such a game unless they are compelled to play it. These mechanisms are therefore ones that cannot emerge through private, one-shot contracts, even if the beliefs about the distribution of valuations in the population are common knowledge.

Collective Action Problems in Sequential Decisionmaking

One of the purposes of law is to enable parties to coordinate their efforts with others. In some cases, we care only that our efforts are coordinated, not how they are coordinated. It does not matter whether the legal rule requires everyone to drive on the right-hand side or the left-hand side of the road, only that it requires everyone to drive on the same side. In many other cases, however, there are different ways for parties to coordinate their efforts and it makes a difference which one is chosen. In this section, we focus on one of these problems of coordination, the problem of *network externalities*. This kind of externality arises when the value that a user derives from a good increases with the number of others who also use it. As we noted at the beginning of the chapter, the problems that externalities raise disappear if the affected parties can contract among themselves. We want to examine cases, however, in which such contracting is not possible.

Some products become more valuable if competing products are built according to the same standard. Consumers are better off if they have a television set that can receive the same signal as other televisions. Only when many individuals have televisions built to the same standard will it pay for a station to broadcast programs that these sets can receive. Computer operating systems are another example of such a standard.

The existence of network externalities creates several different collective action problems. In particular, a preexisting standard may be rejected prematurely in favor of a new standard. This is the problem of excess momentum. Alternatively, a standard may be in place too long. This is the problem of excess inertia. Both of these are examples of collective action problems that legal rules may affect in important ways.

A model can illustrate some of the key issues associated with too rapid and too slow standard adoption. Assume that a new competing standard comes along. This standard too is subject to network externalities. We have one group of existing players who have adopted the old standard and cannot switch to the new one. Player 1 chooses whether to adopt the old standard or the new standard. After Player 1's decision, and with knowledge of that decision, Player 2 makes the same decision. We show in Figure 6.9[8] the subgame that takes place after the existing players have already adopted the old standard.

We can solve this game through backwards induction. If Player 1 adopts the old standard, Player 2 will, too. Player 2 gets $9 from the old as opposed to $0 from the new. If Player 1 adopts the new standard, Player 2 will also, as Player 2 gets $10 from the new and $9 from the old. Player 1 therefore faces a choice between old/old with a payoff of $19 and new/new with a payoff of $10, so Player 1 will adopt the old standard and Player 2 will follow suit.

In this game, the interests of the players do not coincide. Player 2 would prefer that Player 1 adopt the new standard, in which case it would also adopt it. Player 1, however, does not take this course because it would suffer a loss from being an early adopter of the new standard. Until Player 2 enters the market, Player 1 has none of the benefits associated with network externalities. Such is the curse of the early adopter.

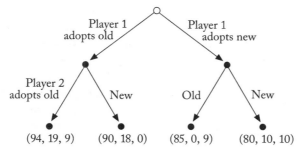

Figure 6.9 Adoption of old or new standard.
Payoffs: Existing Players, Player 1, Player 2.

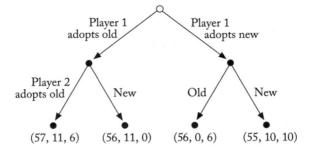

Figure 6.10 Standards and excess inertia. *Payoffs:* Existing Players, Player 1, Player 2.

In this example, the equilibrium is one in which the players act in a way that advances their joint interest. (The payoffs to all the players together in equilibrium are as high as they can be.) Situations can arise, however, in which this is not the case. The game in Figure 6.10 shows a situation in which there is not a match between private decisionmaking and the interests of all the players as a group.

We can again solve this game using backwards induction. Player 2 still follows new with new and old with old. Recognizing this, Player 1 again chooses the old standard and therefore receives a payoff of $11 rather than one of $10. When there are these payoffs, however, the overall social welfare is only $74. If both players adopted the new standard, overall social welfare would be $75.

This game illustrates excess inertia. The old standard should be abandoned by new adopters, but they instead stay with the old standard. Nonetheless, the benefits that accrue to Player 1 are too small relative to the benefits of being on the old standard. Player 1 is in the position of conferring external benefits on either the existing players or the future players, but not necessarily both. If Player 1 adopts the old standard, the existing players will benefit; if Player 1 adopts the new standard, only Player 2 can derive external benefits from Player 1. Player 1 will ignore all the external benefits that accrue to other players.

The game in Figure 6.11 shows that a third outcome is possible as well. As before, Player 2 will play new following new and old following old. Player 1 chooses between old and a payoff of $19 and new and a payoff of $20. Both players therefore embrace the new standard and overall welfare is $121. Overall welfare would have been $122, however, if Player 1 had chosen to stay with the old standard. In this example, the new standard is adopted too quickly, or there is excess

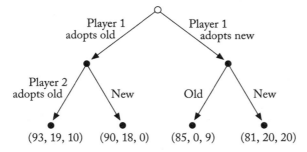

Figure 6.11 Standards and excess momentum. *Payoffs:* Existing Players, Player 1, Player 2.

momentum. Again, Player 1 ignores the external benefits to others. The network externalities of the new standard are large enough so that Player 1 is willing to forgo earning more on the old standard at the outset. These externalities, however, are not sufficiently large to make up for the benefit that would have been conferred on the existing players had both players adopted the old standard. One could, of course, adjust the parameters once again and confront a case in which both players embrace the new standard and that outcome is the one that is socially most beneficial.

These examples convey the sense in which standards can be adopted too quickly or too slowly. As we embed these games in richer settings, other issues arise which may—or may not—help to temper these forces. For example, it matters whether laws recognize a property interest in the standard that allows one person to control it. Such a property interest affects the resources that are invested in creating a standard. A new standard will displace an old, entrenched one if its benefits are large enough. The control that one person has over the standard may allow that person to charge low prices in the first period and offset them by higher prices in a later period. If no one has rights over the standard, those who bear the costs of entering early have no way to recover their losses when competitors enter in a later period. Moreover, such property rights may give the owner an incentive to invest additional resources in creating the standard.

The idea of network externalities and their consequences for standard adoption has substantial implications for questions of antitrust policy. If firms in an industry can work together on creating standards, they may be able to overcome some of the problems they create. Allowing firms to reach agreements on the products they produce, however, brings with it the potential for them to engage in activities that limit competition. As we showed in the last chapter, conditions may be such that creating devices that allow firms to exchange information may lead to anticompetitive conduct even when there is no explicit agreement.

The government can also attempt to overcome the potential costs of network externalities by propagating standards (as it does, for example, in the broadcast industry), but whether this approach proves successful turns again on whether the government itself possesses the information necessary to make the decisions that advance society's interests as a whole.

Network externalities must also be taken into account in determining the appropriate scope of intellectual property protection for computer software. There is general agreement that the computer source and object codes that run the hardware are expressions and hence are subject to copyright protection.[9] What is less clear is whether the nonliteral elements of a program, such as its structure, sequence and organization, or its user-interface, are protected. The extent to which nonliteral elements are copyrightable affects the network externality problem. Protecting a standard may lead to an initial standard's becoming entrenched. In products not characterized by network externalities, a new entrant or a new purchaser of computer software will simply compare the programs on their merits. The consumer will evaluate the two command structures and choose the superior product. The old program will have no special advantage over the new program.

In products with network externalities, the size of the installed customer base matters a great deal, and, as the formal analysis suggested above, a consumer may reject a new, superior product because a network already exists for the old one. If the new product can take advantage of the network elements of the old program, however, consumers may be willing to shift to the newer and superior product.

A recent case presents this possibility. In *Lotus Development Corp. v. Borland International, Inc.*,[10] Lotus alleged that Borland had infringed copyrights for the spreadsheet Lotus 1–2–3. Lotus has long been the dominant spreadsheet for MS-DOS machines. Borland markets Quattro Pro, a competing spreadsheet. Although the case raises many issues, the central issue presented was whether Quattro Pro could use the same command set used by Lotus. The command set consists of the names of particular actions ("Move," "Print," "Copy") and the like, their arrangement in menus and submenus, and the keystrokes for operating the program.

Borland, in short, wanted to copy the user-interface for controlling the source code. It was the part of the program that consumers knew and that may therefore have generated network externalities. Borland did not copy the source code, which, because it worked invisibly to the users, generated no network externalities. The court held that, as a matter of law, Borland had infringed the Lotus copyright by copying much of Lotus's menu commands and menu structure.

Under traditional copyright law, Borland was entitled to copy Lotus's "ideas," but not its "expression." The court found that the particular menu commands and structure used by 1–2–3 were expressions of an idea, whereas the use of menus and submenus represented an uncopyrightable idea. Decisive in the court's view was the wide range of possible ways in which Borland could have presented a user-interface without copying Lotus.

This analysis misses an important point, however. Once Lotus chose a particular user-interface, the network externality effect may shrink the range of competing expressions to zero in the sense that no consumer would buy a competing program, even if it had a superior source code and worked more quickly or used less memory. To focus on the literal number of expressions still available has little meaning. Once the standard is in place and an installed base exists, a competing program may be possible only if it incorporates substantial elements of the existing standard.

One does not want to press this point too strongly or suggest that only the source code itself should give rise to copyright protection. Among many other things, one should take account of the trade-off between allowing and limiting copying. By doing the former, we may mitigate the problem of network externalities, but by doing the latter we may ensure that the incentives exist that are needed to produce efficient standards in the first place.

One must also take into account the scope of the patent laws. Patent law protects ideas themselves, whereas copyright is limited to the particular expression of ideas. To understand the way in which computer software is protected, one must understand the reach of patent law as well as copyright law. Nevertheless, the presence of network externalities in the case of computer software, but not in other copyrighted works such as novels or movies, requires us to engage in a different kind of inquiry.[11]

Herd Behavior

We close this chapter where we started: the collective action problems associated with information and, in particular, the problem of information aggregation. Consider the following situation.[12] Potential employees are once again concerned about both the cash wage offered and the level of risk associated with working for particular employers. Assume that a group of 100 such employees face 2 different employers, Firm and Manufacturer. Each firm offers the same cash wage of $10, so an employee would prefer to work for the safer firm.

There is a known prior probability of 51 percent that Firm is safer than Manufacturer, but each employee receives a private signal about the level of safety associated with the 2 potential employers. And each employee knows that the other potential employees also receive signals. Each employee can observe the actions that other potential employees take in response to these signals. All the signals are of the same strength, and they are independent of one another. If the private information suggests that Manufacturer is safer, an employee would alter the prior belief that Firm is a little safer. Assume further that 99 of the potential employees receive private signals that, contrary to their prior beliefs, Manufacturer is safe. One potential employee receives a signal that Firm is safer. Now consider sequential decisionmaking by the potential employees.

Suppose that the person receiving the signal that Firm is safer chooses first. This person—Player 1—will choose Firm. Player 1's prior belief was that Firm was safer, and the private signal strengthens this belief. Now consider Player 2's decision. The signal that Manufacturer was the safer firm, standing alone, would suffice to overcome the weak prior belief that Firm was safer. (By assumption the signal is strong enough to change the prior belief that Firm is a little safer than Manufacturer.) Player 1's choice, however, has altered Player 2's information.

Player 2 knows that Player 1 would have chosen Manufacturer had Player 1 received a private signal favoring Manufacturer. Instead, Player 1's choice means that Player 1 received a private signal favoring Firm. Player 2 infers Player 1's information from Player 1's choice. Player 2 therefore has 2 signals in hand: the original private signal favoring Manufacturer and the signal derived from Player 1's decision, favoring Firm. The 2 signals cancel out, and Player 2 is left only with the prior belief that Firm is safer. Hence, Player 2, like Player 1, goes to work at Firm.

Now consider the decision of Player 3. Player 3 will know that Players 1 and 2 have chosen Firm. This is consistent with 2 possibilities: both players received signals favoring Firm, or Player 1 received a signal favoring Firm and Player 2 received a signal favoring Manufacturer. Put differently, Player 3 can infer only that Player 1 received a signal favoring Firm. It can infer nothing about what Player 2 learned. Given this, the signal sent to Player 2 is ignored and the one sent to Player 1 is again sufficient to cancel out the signal sent to Player 3. Hence, Player 3 will also ignore the private signal favoring Manufacturer and will select Firm, as will all the subsequent players. All the players except Player 1 ignore their private information, even though each piece of information, standing alone, would be enough to change their actions, and even though the private signals given to the group as a whole overwhelmingly suggest that Manufacturer is the safer firm.

The strategy space in this model is limited. A richer model would, for example, take account of the ways in which Manufacturer might be able to signal that it is in fact the safer firm. It might, for example, find some third party to certify that it is safe in the way we suggested at the beginning of the chapter. Nonetheless, the model may capture an information externality that may lead to substantial collective action problems. In this herd behavior model, those who make the first decisions ignore the effects of their decisions on those who choose later. This result arises because the players act in sequence, and subsequent players ignore the private information they receive. The resulting herd externality leads the group far from where it should be.

This model captures the intuition that sequential decisionmaking may have a momentum that runs contrary to the interests of the group. The sequential, individual decisionmaking may do a poor job of aggregating the information available to each of the players. A more direct mechanism for aggregating the individual information might improve matters. For instance, a union of workers in a particular trade might collect the bits of information the individual members hold and make it generally available. Even if the information could not be gathered, some voting mechanism might be able to aggregate the private information. We see similar devices in other areas of the law. For example, a bankruptcy proceeding might serve as such an aggregation mechanism. Similarly situated creditors vote on such questions as whether to reorganize a firm whose prospects are not clear and about which the creditors have different private, nonverifiable information. This kind of voting rule may give creditors a rule they would bargain for in advance. It may provide a mechanism that processes these signals efficiently and avoids the costs associated with herd behavior.

In addition to the direct aggregation of private information, mechanisms might be devised that prevent inferences from being drawn and break the information chain. Individuals as a group would choose such a mechanism at the outset if they could. Everyone might be better off if each individual decided on the basis of the private information the individual received. If, in our model, Player 2 could not observe Player 1's choice, Player 2 would decide to work for Manufacturer. Player 3 would then observe the choices of Players 1 and 2. Player 3 would know that Player 1 had received information that suggested that Firm was safer and that Player 2 had received information that suggested that Manufacturer was safer. Given Player 3's private signal that favored Manufacturer, Player 3 would also select Manufacturer. Once Player 1's choice is hidden, the adverse inference chain is broken.

Similar problems may arise on stock exchanges and on commodities markets. Traders each possess nonverifiable information about the stock or commodity that they are buying or selling, and this private information affects the price at which they trade. They also attempt to glean information from the trades that others make. Here again, the trade of the first player may be decisive, just as it was for the employees in our example. The actions of the first trader affect the decision of the second trader and all who follow. Subsequent traders can infer only what the first trader's private information must have been. They can infer nothing about the private information of any of the subsequent traders.

In well-developed markets, devices may emerge that interrupt the inference chain. The commodities markets commonly limit price changes to a daily maximum amount. For example, the Chicago Board of Trade limits the daily movement on commodity contracts. On any given day, soybeans may go limit up or limit down, but no more, notwithstanding the fact that the individual traders would like to continue to change the price. In the aftermath of the 1987 Crash, the New York Stock Exchange adopted a series of circuit breakers designed to prevent major movements of the market. A 50-point drop in the Dow Jones Industrial Average or a 12-point drop in the Standard and Poor 500 Index triggers limits on program trading. A 250-point fall of the Dow triggers a trading halt of 1 hour, and an additional fall of 150 points halts trading for another 2 hours. In markets as fluid as those of the major exchanges, these are significant interruptions in trade.

All of these devices prevent satisfaction of trading preferences; they interfere with the momentarily desired private ordering. The herd behavior model suggests that these limits might enhance overall welfare and that the limits might therefore be agreed to voluntarily. We do see such devices in our best-organized markets, but we cannot infer whether these devices were adopted voluntarily or in response to government pressure. Limits on daily moves of commodity prices were first adopted during World War I, after the cotton market had dropped precipitously following an announcement by the German government that it intended to sink all ships in certain zones.[13] The federal government asked the exchanges to impose limits, and, not surprisingly, they did so. The NYSE adopted its circuit breakers in the face of substantial pressure for more direct regulation. If we were to see our best-organized markets adopt such devices entirely on their own initiatives, however, it would cause one to consider whether legal rules might appropriately introduce them in other situations in which they might not naturally arise, but would be beneficial nevertheless.

Summary

In this chapter, we have explored two kinds of complications that we need to take into account in modeling strategic behavior. Legal analysts and lawmakers must necessarily focus their attention and separate the problem at hand from larger ones. Nevertheless, one must always keep the larger picture in mind. A strategic interaction between individuals may have a different outcome and a different structure once it is put into its larger context. What appears to be a prisoner's dilemma problem that the law must address may be nothing of the sort once one understands the context in which the problem arises. In addition, although it is often useful to reduce complicated problems to two-person games, we must also be mindful that some problems, such as those of network externalities and herd behavior, are necessarily ones that arise when multiple parties are involved and hence resist such simplification.

Bibliographic Notes

Introduction to collective action problems. Olson (1965) and Hardin (1968) are basic introductions to the problem of collective action. For discussions of the issues raised by OSHA, see Posner (1992, pp. 334–335) and Sunstein (1990, pp. 48–55).

Embedded games. For work in political science regarding the risk of misidentifying the game being played, see Tsebelis (1990); see also Ostrom (1990, pp. 8–18). The form of the game in Figure 6.2 is taken from Kohlberg (1990), where it plays an important role in demonstrating the usefulness of "forward induction" arguments in solving games. For other work on forward induction, see van Damme (1989) and Ben-Porath and Dekel (1992). The view that the failing firm should be analogized to a common pool and that bankruptcy should be seen as a response to that problem was originally developed in Jackson (1982) and Baird and Jackson (1984); see also Jackson (1986). Our discussion of embedding the bankruptcy problem into decisions over the firm's capital structure follows Picker (1992).

Collective action and private information. A good technical textbook on public economics is Laffont (1988). The Clarke-Groves mechanism was independently developed by Groves (1973) and Clarke (1971). They are closely related to Vickrey (1961). The budget-balancing mechanism is derived from d'Aspremont and Gerard-Varet (1979) and Arrow (1979). A good introduction to and survey of the vast literature on incentive theory is Laffont and Maskin (1982). For an introduction to mechanism design that discusses its relevance to law, see Emons (1993).

Collective action problems in sequential decisionmaking. For general treatments of network externalities, see Katz and Shapiro (1985), (1986) and Farrell and Saloner (1985), (1986). Menell (1987), (1989) applies these ideas to computer software and copyrights; see also Farrell (1989). For early work on the legal consequences of network effects, see Carlton and Klamer (1983).

Herd behavior. Our discussion of the economics of herd behavior follows Banerjee (1992).

CHAPTER 7

Noncooperative Bargaining

Modeling the Division of Gains from Trade

In the first six chapters of this book we have seen how legal rules provide parties with a backdrop against which all their bargaining takes place. We have concerned ourselves primarily with the way in which different legal rules affect the substantive bargains themselves—such as whether goods were sold with a warranty or whether a carrier was liable for delays in transporting goods. In this chapter, we explore the way in which formal models can help us understand both how parties reach agreement and, if they do, how they divide the gains from trade between themselves.

An owner values possessing a book at $10. A buyer will spend as much as $15 to acquire the same book. The different values that the buyer and the seller place on the book make a mutually beneficial trade possible, and legal rules should ensure that parties who want to make this trade are able to do so. After all, such a trade can leave both parties better off. Legal rules, however, do more than simply facilitate trade. They also may affect the way the parties divide the potential gains from any trade (in this case the $5 difference between $10 and $15). Two labor laws might be equally efficient, but one might create a bargaining environment that leaves the workers with higher wages than the other.

Much bargaining takes place between parties who have an established relationship. Workers who bargain over a new contract may have worked at the same firm for many years and may have developed substantial firm-specific skills. These workers have valuable skills that the employer cannot find in new employees, but which have no value to the workers in any other job. The rules governing the contract negotiations determine how the extra value that the workers bring to the firm is divided. These rules in turn affect the incentives of the parties during the period of the original contract. If the bargaining environment is one in which the firm enjoys the entire surplus, the workers will have no incentive to develop firm-specific skills. If, however, the workers enjoy the entire surplus, the firm will have no incentive to spend resources training its workers and giving them firm-specific skills. As we first saw in our discussion of renegotiations in Chapter 3, the rules governing such negotiations matter because they affect how people behave before the negotiations take place.

The need to ensure that the rules governing bargaining give parties the right set of incentives throughout their relationship permeates the law. As we have seen, parties to a contract may be able to structure the environment in which future bargaining takes place at the time of their initial contract. The law, however, must still supply a set of default rules. In addition, there may be other environments, such as bankruptcy, in which bargaining affects the rights of third parties, and the rules cannot be left entirely to the parties themselves. There are still other contexts, such as labor negotiations, in which lawmakers want to affect the way gains from trade are divided quite apart from what the parties would agree upon in a bargain before the fact.

In any particular case, much will turn on the specific facts and the reputations and other characteristics of the parties. One party, for example, may be willing to forgo any benefit from reaching a deal in order to establish a reputation as a tough negotiator. The two individuals may live in a culture in which there are strong norms about how such divisions are to be made. The fear of ostracism may drive them toward a particular division. Our focus, however, is on how laws, as a general matter, affect negotiations; or, to put the point more precisely, we want to know how a change in the legal rules is going to change the bargaining environment in which parties operate.

When we capture the interactions between the players in the simplest way, many bargaining problems we confront are variations on the following game. Two players are seated at a table. In the center of the table is a dollar bill. The players must negotiate with each other and agree on a way of dividing the dollar. One makes an offer that the other can accept or reject. Once a player rejects an offer, that player can make a counteroffer that the first player can in turn accept or reject. When they reach some agreement on how to divide the dollar, they will each receive their respective share of the dollar. Unless and until they reach agreement, however, they receive nothing. Delay, of course, matters because a dollar today is worth more than a dollar at some later time.

These games—ones with alternating offers and infinite horizons— were explored by Ariel Rubinstein in the early 1980s and are often called *Rubinstein bargaining games.* There are, of course, other ways to model bargaining. Indeed, Rubinstein himself offered an alternative model in which the parties are driven toward agreement, not because they want the benefits of reaching agreement sooner rather than later, but because each round of bargaining is costly. The game of splitting a dollar in which each party wants to reach agreement sooner rather than later, however, has advantages over other models that we might use. Its basic solution—one in which similarly situated players divide the dollar evenly between themselves—comports with our intuitions about how people behave. In addition, it is easy to introduce changes in substantive legal rules into the bargaining environment in this model.

A legal rule can be seen as an *exit option,* the right of one party to cut short the give-and-take of bargaining and receive some alternative payoff. An alternating offer game with exit options generates many predictions about how changes in legal rules affect bargains. We cannot assume automatically that these predictions are in fact consistent with the way the world works, but, as elsewhere, they do reveal tensions and problems in legal rules that are not otherwise immediately evident.

Before we introduce exit options, we need to identify the solution to the simple game in which the two players must split a dollar. There are several approaches to solving this game. One is to begin by asking whether there is any equilibrium in which the first player's strategy is to make an offer, to refuse all counteroffers for any lesser amount, and to keep repeating the initial offer until it is accepted. (It seems logical to ask if such a strategy exists because this game is one of full information and a player does not learn anything during the course of the game. When the first player makes an offer on any move, the player is in the same position the player was in at the start of the game, except for the passage of time.)

Such a strategy can be part of a subgame perfect equilibrium only if the second player cannot respond with a counteroffer that the first player is better off taking rather than repeating the initial offer. If the second player can make such an offer, then the first player's strategy of repeating the initial offer cannot be part of a subgame perfect equilibrium. Such an equilibrium can exist only if the first player chooses a best response in the subgame that exists after the second player makes the counteroffer. The strategy of not accepting any counteroffer is a best response only if the first player is better off repeating the initial offer than taking the counteroffer. If the offer of the first player leaves the second player with too little, the second player is able to make a counteroffer that the first player would prefer to repeating the initial offer. It turns out that the first player's desire to take as large a share as possible and the second player's ability to make an attractive counteroffer enable us to identify a unique solution to this game.

Let us start by asking whether an equilibrium can exist in which the first player's strategy is to offer the second player a share of the $1 equal to s on every move. To assess this strategy, we must determine how much the value of s shrinks if the players wait a period before agreeing on a division. As in Chapter 5, we can use δ to represent the amount that the value of the dollar decreases for a party during each period. (For example, if δ is 0.5, then that party values receiving money in the first period twice as much as receiving the same amount of money in the next period. A party is indifferent between receiving $0.50 in the first period and $1 in the second.) To keep things simple, we assume at the outset that both players have equal discount rates.

When a player repeats an offer rather than accepting a counteroffer, the first player must wait at least an additional period before receiving a payoff. Hence, the first player's best response is to accept a counteroffer, even if it is less than the initial offer, if it gives the first player more than the initial offer would give when accepted one period later. By offering s, the first player retains a share of the dollar equal to $1 - s$. Because the first player must wait an additional period by repeating the offer, however, a counteroffer from the second player is more attractive if it gives the first player more than $(1 - s)\, \delta$.

It will be a best response for the second player to make such a counteroffer, however, only if making it leaves the second player better off than taking the first player's initial offer. Because the second player must wait an additional period, the counteroffer must leave the second player with an amount larger than s. More precisely, it must leave the second player with an amount that is greater than s/δ. Therefore, the second player can never offer the first player more than $1 - s/\delta$, but it is in the second player's interest to make a counteroffer of $(1 - s)\, \delta$. The first player is better off taking such a counteroffer rather than repeating the initial offer.

An equilibrium in which the first player repeats the initial offer and never accepts a counteroffer cannot exist if (1) the second player is able to make a counteroffer of $(1 - s)\, \delta$, and (2) this amount is greater than

$1 - s/\delta$. The first player can ensure that the second player cannot make such a counteroffer by making an initial offer such that:

$$1 - \frac{s}{\delta} = (1 - s)\,\delta.$$

This equation is straightforward. When the first player chooses a value of s that satisfies this equation, the second player has nothing to gain from making a counteroffer. The only counteroffer that the first player would be better off accepting rather than playing the equilibrium strategy leaves the second player with no more than the second player would receive by taking the first player's offer. When we do the algebra to find out the share that the first player offers the second player at the outset, we find that:

$$s = \frac{\delta}{1 + \delta}.$$

We subtract this amount from \$1 to find out what share the first player would keep.

At this point, we have discovered a subgame perfect equilibrium to this game. The first player offers the second player $\delta / (1 + \delta)$ and does not accept any counteroffer. The second player accepts any offer equal to or greater than $\delta / (1 + \delta)$ and otherwise makes a counteroffer of $(1 - s)\,\delta$. The first player receives $1 - s$ or $1 / (1 + \delta)$, which is the most that the first player can receive, given the strategy of the second player. Never accepting a counteroffer is a best response to the strategy of the second player, given that the second player does not make a counteroffer larger than $(1 - s)\,\delta$. The strategy of the second player is also a best response. Given the strategy of the first player, the second player's best response is to take the initial offer.

There is no other combination of strategies in which the first player makes an initial offer and never changes it that is also a subgame perfect equilibrium. The ability of the second player to make a counteroffer that the first player would be better off taking drives this result. The first player makes an offer that is just large enough so that the second player is not able to make a counteroffer that would prevent the strategy of repeating the same offer from being a best response. Indeed, this combination of strategies turns out to be the only subgame perfect equilibrium to this game.[1]

An only slightly more complicated version of the Rubinstein bargaining model posits different discount rates for the parties (δ_1 and δ_2 respectively). When the time period between offers becomes arbitrarily short, the share that the second player enjoys is:

$$\frac{\ln \delta_1}{\ln \delta_1 + \ln \delta_2}.$$

If the two discount rates are the same, the share that the first player offers the second in the first round approaches $\frac{1}{2}$ as bargaining periods become arbitrarily short.[2] This comports with the intuition that, when everything else is equal, parties who bargain with each other will tend to split the difference.

Legal Rules as Exit Options

In this section, we use the alternating offers model to examine different legal rules and the way they affect the kinds of bargains parties strike. In the simple Rubinstein bargaining game, the parties do not have any alternative to striking a deal with each other; the players receive nothing unless they reach an agreement with each other. In most actual bargains, however, what matters is the alternatives that each of the parties enjoy. Many legal rules do not affect the actual bargaining process itself, but rather the alternatives that each party has to continuing the negotiations. The bargaining problem in *Peevyhouse v. Garland Coal and Mining Co.* provides a good illustration of how legal rules can affect the negotiations between the parties.[3]

As we saw when we first discussed the case in Chapter 4, Garland broke its promise to restore the land to its original condition when it finished strip mining. To highlight the bargaining problem, we make a number of simplifying assumptions. We assume that the land in its current condition is worthless, but that it would cost Garland $1 million to restore the land. We begin by assuming that the value that the Peevyhouses attach to their land is $800,000 and that this value is observable but not verifiable. It is plausible to think that the information is nonverifiable. A farm is not a fungible commodity like wheat or corn. It may not have a readily ascertainable market value. The farm may be worth more to the Peevyhouses than to anyone else, and a court may have no way of knowing how much more. The value that the Peevyhouses attach to the land, however, may be known to Garland because of its earlier negotiations with the Peevyhouses.

One of the goals of contract law is to give the party in the position of Garland the right set of incentives throughout the course of the contract. Garland should make its decisions taking full account of the loss that the Peevyhouses suffer if the land is not restored. When transaction costs prevent the parties from opting out of any default rule, we want the default rule governing the Peevyhouses' remedy against Garland to be such that Garland has to pay the Peevyhouses an amount equal to the harm that the Peevyhouses suffer from the breach. No other amount ensures that Garland takes account of the Peevyhouses' interests while it is performing the contract. Creating such a default rule, however, is hard when the amount of harm that the Peevyhouses suffer is observable but not verifiable information.

One possibility is to give the Peevyhouses a right to specific performance. If Garland is forced to restore the land, there will be economic waste. By assumption, the land is worth less to the Peevyhouses than the cost of restoring it. Both the Peevyhouses and Garland are better off if the land is not restored. By ordering Garland to restore the land, the court may bring about an inefficient outcome if renegotiations are not possible. Garland and the Peevyhouses, however, may be able to bargain with each other before the Peevyhouses actually invoke their right to specific performance. We need to take account of the dynamics of these negotiations before we can properly assess the merits of a specific performance remedy in this case. We need to consider whether the bargaining between the parties proceeds in such a way that the land is not actually restored, but Garland offers and the Peevyhouses accept an amount of money equal to the subjective value that the Peevyhouses place on the land.

We know that Garland must offer an amount that is at least as large as the value that the Peevyhouses attach to having the land restored. (If it made an offer for any less, the Peevyhouses would reject it and insist upon specific performance.) If the parties settle on an amount larger than $800,000, however, Garland has the wrong set of incentives. It may spend money during the course of the contract as if the harm the Peeveyhouses suffer from breach is greater than it is. But it is possible that the parties will in fact settle on an amount that is equal to $800,000. We can see this by extending the alternating offers model we set out in the last section.

To put this problem in its simplest form, let us assume that judgment has already been entered against Garland and it has had to post a bond of $1 million, the amount equal to the cost of restoring the land. The bond will remain in effect until the Peevyhouses insist on specific performance and Garland is forced to restore the land or until the Peevyhouses and Garland reach a settlement. The Peevyhouses, of course, would like to have their land restored or their cash settlement sooner rather than later. Garland, for its part, would like to reach an agreement and be able to enjoy some part of the money that is now tied up in the bond. The Peevyhouses can make an offer or cease negotiations and demand specific performance. Garland can either accept the offer or make a counteroffer. The Peevyhouses can then either accept the counteroffer, make another offer, or abandon the negotiations and demand specific performance.

We shall assume that once the Peevyhouses demand specific performance, the bargaining cannot begin again. These negotiations are a Rubinstein bargaining game, except that one of the players (the Peevyhouses) has the right to walk away from the bargaining table. This right to walk away—this exit option—is something we must take into account in solving the game.

We begin by positing that a stationary equilibrium exists similar to the one we saw in the original game. There are two possibilities to consider—the value of the exit option may be greater than the offer the Peevyhouses would make if they were playing a simple Rubinstein bargaining game (an amount we shall call their *bargained-for share*) or it may be less. Let us assume first that the subjective value that the Peevyhouses place on the land once restored is only $200,000. In this event, having the exit option, having the ability to force Garland to restore the land, may do the Peevyhouses no good. They are better off playing the Rubinstein bargaining game, ignoring their exit option, and receiving what they would receive in that game.

We know from our solution to the original game that, if the Peevyhouses and Garland had the same discount rate and if the time between offers were short, the Peevyhouses could ask for $500,000 at the start of negotiations and Garland would give it to them, even if the Peevyhouses had no exit option. If the value of the exit option is sufficiently low, the exit option itself does not affect the play of the game. The exit option does not give the Peevyhouses a way to get more than what they could get without it. The threat to exercise the exit option is not credible, because the Peevyhouses are better off playing the Rubinstein bargaining game than exercising their exit option.

This model suggests that granting the Peevyhouses a right to specific performance in such a case gives Garland the wrong set of incentives during the course of performing the contract. Garland would make investments to avoid breach in order not to pay the Peevyhouses $500,000. Some of these precautions would be wasteful, however, because the Peevyhouses would suffer only $200,000 in damages in the event of breach.

The subjective value that the Peevyhouses place on the land, however, may be high relative to the cost of restoring the land. Let us assume it is $800,000. In this case, the problem is more complicated. The Peevyhouses will never settle for less than the value to them of having the land restored, for the Peevyhouses can receive the equivalent of this amount by exiting from the bargaining and demanding specific performance. Hence, if the Peevyhouses made an offer of $800,000, an amount equal to their exit option, Garland would accept it because it could never hope to do better. By accepting the offer, Garland would leave itself with the most it could ever receive. The question therefore is whether Garland would ever accept an offer that was any higher than the $800,000 subjective value the Peevyhouses put on the land. (We should bear in mind that, under our assumptions here, Garland knows this value.)

To answer this question, consider a proposed equilibrium in which the Peevyhouses choose a strategy where they insist on a given amount greater than their exit option and repeat this offer every period. Would Garland ever accept such an offer, rather than make a counteroffer? In this proposed equilibrium, the offer the Peevyhouses make in their first move is the same as the one they make in all subsequent rounds of the game. In this proposed equilibrium, however, Garland will always be able to make a counteroffer that is more attractive to the Peevyhouses than repeating their initial offer. Garland, in other words, can make a counteroffer that leaves it with more than the Peevyhouses offered initially, taking into account the delay. We can also show that this counteroffer is better for the Peevyhouses than their original offer would be to them if it were repeated and then accepted. The proposed strategy of repeating the initial offer is not a best response and is therefore not part of an equilibrium.

In the absence of exit options, the Peevyhouses would receive their bargained-for share. The value of the exit option ($800,000) is higher than the amount of the bargained-for share ($500,000 in this example). Any offer that is higher than the exit option must be higher than the bargained-for share as well. Let us assume that the Peevyhouses demand $900,000. In the face of any such demand, Garland will always be able to make an attractive counteroffer. Garland can offer less than $900,000 and the Peevyhouses will be better off taking it than repeating the offer of $900,000. The Peevyhouses are better off taking the counteroffer once the effects of discounting are taken into account. As we saw in the previous section, whenever the Peevyhouses demand to keep more than $500,000, Garland can make a counteroffer that would leave the Peevyhouses better off than they would be if they repeated the same offer again.

Because Garland always has the ability to make an attractive counteroffer, it cannot be an equilibrium for the Peevyhouses to adopt a strategy of repeating a demand greater than the value of the exit option at every move. Nor can they demand some amount greater than the value of the exit option in one period by threatening to demand even more in a subsequent period. Such a threat is not credible. The alternating offer game with an exit option is different only in that the player with the exit option has the ability to exit the bargaining and this ability puts a floor on what that player receives in any bargain. As the time intervals become sufficiently short, the only subgame perfect equilibrium is one in which the Peevyhouses offer an amount that is equal to the value of their exit option in every period and Garland accepts the offer in the first period.[4]

When the game is one of observable, nonverifiable information, the players do one of two things. When the value of the exit option is below the bargained-for share, each player receives the bargained-for share of what is at stake—the amount that Garland will spend on restoring the land. When the Peevyhouses' exit option is high enough, however, the shares are no longer driven by the dynamics of bargaining but instead by exactly the thing that we want to matter, the subjective damages the Peevyhouses have suffered as a result of Garland's failure to keep its promises.

This model suggests that giving the Peevyhouses a specific performance remedy might give Garland a better set of incentives during the course of the contract than requiring a court to award damages. The postbreach negotiations between the parties prevent specific performance from ever taking place, and the amount on which the parties settle may be exactly equal to the subjective value that the Peevyhouses place upon the land. By threatening to order specific performance, in other words, a court might induce a bargaining process that would lead to a payment from Garland to Peevyhouse that was exactly equal to what the court would award in damages if it possessed the relevant information.

Up to this point, we have assumed that the subjective value that the Peevyhouses place on the land is observable information. It may be the case, however, that Garland, like the court, may not know this information. We can again consider the two cases. If the value of the exit option is less than the share that the Peevyhouses would receive from bargaining, the Peevyhouses would be content with their bargained-for share, except to the extent that they can make Garland believe that their subjective value is in fact high. If the Peevyhouses have a subjective value that is higher than the bargained-for share, however, they would seek a way of credibly communicating their higher value to Garland.

Let us begin with a model in which the Peevyhouses place a value of either $200,000 or $800,000 on the land. The Peevyhouses can either exit the bargaining and have the land restored or make an offer. If Garland rejects the offer, a period of time passes before Garland makes a counteroffer. The Peevyhouses can either accept the counteroffer or wait a period of time and then either exit or make another offer. Garland does not know whether the Peevyhouses have an $800,000 or a $200,000 valuation, but it believes that each is equally likely.

Let us assume first that the Peevyhouses place a value of $800,000 on having the land restored. The Peevyhouses should never be able to recover more than $800,000. We know from the previous section that Garland would never agree to pay more even if it knew that the Peevyhouses' exit option was worth $800,000. Garland should therefore never agree to pay more when it knows only that the value cannot exceed $800,000, but might be much less.

The next possibility is that the Peevyhouses can make an offer that reserves $800,000 for themselves when they have the high valuation. Would Garland accept such an offer? There are two possibilities. The first is a separating equilibrium in which the Peevyhouses with the low valuation make a low offer, the Peevyhouses with a high valuation make a high offer, and Garland believes that those who make high offers are high-value and that those who make low offers are low-value. We can eliminate this equilibrium. Given Garland's beliefs, the Peevyhouses' best response when they have a low valuation is to make a high offer. In this proposed equilibrium, Garland believes that the Peevyhouses that reserve $800,000 for themselves have a valuation of $800,000. The Peevyhouses with a valuation of $200,000 could therefore mimic the actions of those with the higher valuation if Garland has this belief. Hence, the combination of actions and beliefs in which the different types separate themselves cannot be an equilibrium.

The second possibility is a pooling equilibrium in which both high- and low-value Peevyhouses make offers in which they reserve $800,000 and Garland accepts. This proposed equilibrium is also one in which actions and beliefs are not consistent with each other. Garland is not choosing a best response given the actions of the Peeveyhouses. If Garland accepts the offer, it recovers $200,000. By turning down the offer and making a counteroffer of $500,000, it stands to do better. The high-value Peevyhouses would reject such an offer because they can get something (their land restored) that they value at the equivalent of $800,000 by exiting. Thus, they prefer to exit rather than to make another offer. Once they exit, however, only the low-value Peevyhouses are left. Garland can infer their type, and the game becomes one that is identical to that in which the information is observable. In such a game, both parties will settle for their bargained-for shares of $500,000 each. By making this counteroffer, Garland expects to recover $250,000 of the million dollar bond it posted.[5] This is better than recovering only $200,000, as it would if it accepted the Peevyhouses' offer at the outset. Because Garland is better off by making a counteroffer, there cannot be an equilibrium in which Garland accepts offers from both that leave it with $200,000.

We have eliminated all the potential equilibria in which the Peevyhouses with the high valuation make an offer of $800,000 and that offer is accepted. Hence, the Peevyhouses with the high valuation must decide at their first move whether to make an offer of less than $800,000 or to exercise their right to exit. Given that no offer will be accepted that will give them as much as exiting immediately, they will exit. Because they exit, Garland infers the type of the remaining players and the game becomes the same as one of observable information. The Peevyhouses with the low valuation simply receive their bargained-for share.

Versions of this game in which the Peevyhouses might have many different valuations are harder to solve, but the solutions to these games have the same general features as the one to this game. Peevyhouses with high valuations exercise their exit option. They have nothing to gain from bargaining in a world in which they cannot readily distinguish themselves from those with lower valuations. If they never receive more than the value of the exit option by continuing in the game, they exit.

This analysis suggests that some unraveling may take place. The Peevyhouses may invoke their right to specific performance when their subjective values are high enough. The same force that makes specific performance attractive when the information is observable—the inability of the first player to obtain more than the value of the exit option—leads to a bargaining breakdown when the subjective value of having the farm restored is nonobservable, nonverifiable information. The first player will not incur the costs of bargaining if nothing can be gained by it.

The possibility of bargaining failure compounds another difficulty that exists if the subjective value of the farm to the Peevyhouses is private, nonverifiable information. The purpose of awarding the Peevyhouses these damages is to induce Garland to internalize the costs that it imposes on the Peevyhouses if it fails to restore the land. Even if this bargaining led to an outcome in which the Peevyhouses were given the subjective value, it would fail to give Garland the correct set of incentives. Garland needs to know the subjective value that the Peevyhouses place on the farm at the outset. If it does not possess this information, it will take too many precautions to prevent default in some situations and not enough in others. The possibility of bargaining failure, coupled with the need to provide information to Garland at the outset, reinforces the need to consider default rules that induce parties to bargain about such matters at the outset.

We do not know how accurately the Rubinstein bargaining model captures the dynamics of real world bargaining. It sheds light, however, on the crucial question: To what extent does an exit option, an ability to leave the bargaining table unilaterally, affect the dynamics of bargaining itself? The model suggests that much turns on whether the subjective value that the Peevyhouses place on having their land restored is private information. If the value is known to both Garland and the court, there is nothing to be gained from having a specific performance award. If the value is known to Garland, but not to the court, there is at least the possibility that bargaining between the parties will lead to their settling on an amount that is in fact equal to the subjective value that the Peevyhouses place on the land. When the value is information that is private to the Peevyhouses, however, there is a possibility that bargaining will regularly break down and that the land will be restored even when it is in no one's interest.

This analysis suggests that one can incorporate legal rules into a bargaining environment by modeling them as exit options, as rights available to a player when no consensual bargain is reached. There are many rules in contract law whose principal effect may be to create a context in which parties negotiate. A buyer, for example, has a right to *perfect tender*. A buyer may reject goods and force a seller to take them back if they do not conform to the contract. When reshipping the goods is costly or when they are suited to a buyer's special needs, the buyer may be the person who should end up with the goods even though they are nonconforming. In such a case, the buyer's right to perfect tender affects not whether the buyer ends up with the goods, but rather how much of a discount the seller must offer to persuade the buyer to keep them.

As in *Peevyhouse*, the effect of the perfect tender rule is two-fold; it changes the incentives of one of the parties before the fact (in this case, the incentive of the seller to ensure that goods are conforming in the first instance), and it sets the initial conditions for bargaining between the parties. Here again, a Rubinstein bargaining model can shed light on such questions as how the bargaining between the parties will change when the value that the buyer attaches to the goods is observable by the other party, but not by the court, or when the value is known only to the buyer.

The model we have developed in this section might also be used to explore the parallel question of *substantial performance* in contract law. A builder finishes a building but installs the wrong kind of pipe. The pipe has little or no effect on the value of the building, but it would cost a great deal to take out the pipe and install the correct kind. Can the builder still sue for the balance of what it is owed or does the builder's breach give the owner the right to withhold payment? Again, if it makes little economic sense to replace the pipe, the legal rule will only determine the nature of the negotiations between the builder and the owner. The traditional legal rule, one that gives the builder the right to sue in the event that its performance is *substantial* can again be subject to scrutiny using this model. Rather than pursue this or other bargaining problems that arise in contract law, however, we apply the model to a completely different body of law in the next section and explore the light it can shed on the law of corporate reorganizations.

Bargaining and Corporate Reorganizations

When a firm that is worth keeping intact as a *going concern* finds itself unable to pay its debts, it must enter into negotiations with its creditors and restructure its debt. Even if the firm never files a bankruptcy petition, the course that the negotiations take will be shaped by the rights that parties have in bankruptcy. We can use the Rubinstein bargaining model with exit options to understand how different bankruptcy rules affect the rights of the parties.

We begin with a representative example of a closely held firm that finds itself in Chapter 11 and is a good candidate for a successful reorganization. Firm made an effort to expand its operations and borrowed heavily from Creditor in the process. Creditor took a security interest in all the assets of Firm, and therefore has a priority claim on those assets. The effort to expand proved a disaster, as the economy took a turn for the worse at just the wrong time. In the end, Firm had to retrench. The heart of the business may be basically sound, but the amount owed Creditor alone may exceed the value of Firm. Firm is worth more if it is kept intact as a going concern, but only if Manager continues to run it. Manager's skills, contacts with customers, and ability to improve Firm's patented products may make Firm worth much more than if anyone else ran it. Manager's skills are not fungible, and, to have the proper incentives, Manager, like the managers of most closely held firms, must retain an equity interest in Firm.

Bankruptcy law creates a bargaining environment in which Creditor and Manager negotiate with each other. At the time of the loan, Creditor and Manager can anticipate the bargaining environment in which they will find themselves in the event that Firm needs to be reorganized, as well as the division of assets that such a bargaining environment will produce. If Creditor can predict that it is likely to receive a small share in the event of a reorganization, it will demand a correspondingly high interest rate at the time of the initial loan. Full compensation for these risks, however, does not make the division of Firm in a reorganization irrelevant. Firm will be more likely to default if it must pay a higher rate of interest to Creditor (assuming that the debt level is fixed and the probability distribution of Firm's returns remains unchanged). If there are social costs associated with default—and there almost surely are—giving a smaller share of Firm to Creditor in the event of a reorganization may cause welfare losses. There are other effects that also need to be taken into account, including the need to give Manager the right set of incentives in good times as well as in bad.

When a bankruptcy petition is filed, an *automatic stay* goes into effect.[6] This stay prevents any creditor from exercising rights against the debtor on default. These rights typically include the right to go to court and obtain an order allowing the creditor to seize the debtor's assets. While the automatic stay is in effect, the creditors and the old equityholders try to agree upon a plan of reorganization. Creditor has no right to reach any of the assets if Firm has value as a going concern.

We can examine the rules of United States bankruptcy law and some alternatives to them by returning to the Rubinstein bargaining model and treating the rights that Creditor and Manager have as exit options. The effect of the automatic stay, in terms of our model, is to deny Creditor an exit option whenever, as in the example with Firm, the liquidation value of the assets is less than the value of Firm as a going concern. Because no exit option exists, the liquidation value of the assets that Creditor would be able to enjoy outside of bankruptcy is irrelevant. Because Creditor has no ability to lift the automatic stay, Manager need not pay attention to the liquidation value of the assets during the bargaining. The Bankruptcy Code gives Creditor the right to prevent the confirmation of a plan that does not give it the liquidation value of Firm's assets, but this right does it no good in any negotiations with Manager. With or without this right, Creditor has no way to force Manager to pay it the liquidation value of the assets and no way to extricate itself from the bargaining process.

Manager, however, is most likely in a different position. In the terms of our model, Manager enjoys at least one exit option and may in fact enjoy a second. In the kind of case that we are considering, Manager has rarely signed a long-term employment contract with Firm. Even if Manager has signed such a contract, courts will not specifically enforce it and may refuse to enforce a covenant not to compete if it sweeps too broadly. Hence, during negotiations with Creditor, Manager can threaten to leave Firm and find work elsewhere. The amount that Manager can command in some alternative line of work puts a floor on what Creditor will have to give Manager to continue to manage Firm. In bargaining with Creditor, Manager will therefore insist upon receiving at least this amount.

When the assets remain in Firm and Manager continues to work there, a bargain between Creditor and Manager increases the joint welfare of the two parties. Because of Manager's exit option, Creditor can never receive more than the difference between the value of Firm as a going concern with Manager and the value of Manager's alternative wage. The Rubinstein bargaining model would also suggest, however, that Manager's exit option will play a role only when the amount that Manager can earn elsewhere exceeds the bargained-for share, the amount that Manager would receive if it bargained over Firm and had no ability to work elsewhere. In many cases, the value of Manager's alternative wage may be less than the bargained-for share and hence will not figure in the bargaining.

We can also use the idea of incorporating exit options into a Rubinstein bargaining game to examine the current dispute over what is called the *new value exception* to the absolute priority rule. Cases such as *Case v. Los Angeles Lumber Products* suggest that Manager may be able to force a plan of reorganization upon Creditor in which Firm continues as a going concern in Manager's hands. Under such a plan, Creditor receives a bundle of rights worth only the liquidation value of the assets. Manager keeps the residual.[7] The new value exception enables Manager to force Creditor to take a share of Firm equal to the liquidation value of Firm's assets. Manager can capture the difference, the entire going-concern surplus, without reaching a consensual bargain. The model also suggests, however, that the ability to exit the bargaining and receive this amount does not enable Manager to strike a bargain for any more than this difference. Whether such a right exists and what its exact contours are if it does remain unclear.

Those who have argued against the new value exception have focused primarily on the valuation difficulties that necessarily enter the picture. Creditor may not receive even the liquidation value of the assets if it must rely on a bankruptcy court to determine whether its new interest in the reorganized firm is worth what Manager claims. By using a Rubinstein bargaining model with exit options, however, we can focus on an antecedent question: Does Manager's ability to invoke the exception change the way Firm is divided, even if there are no valuation problems?

The new value exception gives Manager a second exit option because, with it, Manager has another way of truncating bargaining with Creditor. Like the ability to leave the firm and earn a wage elsewhere, the new value exception gives Manager an alternative to reaching a deal with Creditor. The new value exception provides Manager with another threat. Manager will never offer Creditor more in the course of negotiations than it would cost to force Creditor out unilaterally under the new value exception. In this case, of course, Creditor is the one who literally "exits" from Firm. Nevertheless, the new value is best modeled as an additional exit option for Manager. What matters is how a legal rule gives one party or the other a credible threat. In this sense, any ability to cut the bargaining short is an exit option, because it puts a floor on what a party will insist upon in any bargaining.

When Manager has a new value exit option and when Firm has value as a going concern, this exit option—this way of threatening to terminate the bargaining—dominates the ability to earn a wage elsewhere. If the liquidation value is low relative to the bargained-for share, Manager will exercise the new value right. As the liquidation value increases relative to the bargained-for share, increases in the liquidation value directly increase the payments to Creditor and reduce those to

Manager. Once the liquidation value exceeds the bargained-for share, however, Manager is better off threatening to continue to bargain with Creditor than threatening to pay it the liquidation value of the assets. Both Creditor and Manager will receive their bargained-for share. Further increases in the liquidation value relative to the value of Firm as a going concern have no effect on the distributions. Note that in either case, Manager will always receive at least the entire going-concern surplus.

We can contrast bargaining in this environment with that which would take place if Creditor did have the power to declare a default and seize the assets. Canadian law gives Creditor such a right several months after the filing of the petition. We would again have the basic Rubinstein bargaining model, but, in this case, both parties would enjoy exit options. When Creditor and Manager both assess their exit options, they will compare them with the value of their bargained-for shares.

As we have seen, the exit options do not themselves affect the size of the bargained-for share. For example, when Manager will do better by insisting on a bargained-for share than by exiting (by capturing the going-concern surplus or by taking another job), the threat to exit is not credible and will not be a factor in the bargaining. If Manager's alternative wage is less than Manager's bargained-for share and Firm enjoys a going-concern surplus, two factors will determine the division of Firm in a reorganization: (1) whether Manager has a new value exit option; and (2) the relationship between the liquidation value of the assets and Creditor's bargained-for share of Firm as a going concern.

By contrast, if Creditor did have an exit option (that is, if it were able to reach its collateral and sell it), the division of Firm would be quite different. The going-concern surplus that Manager can capture using the new value exception and the value of Manager's alternative wage may often be less than Manager's bargained-for share. In such cases, giving Creditor an exit option has a dramatic effect on the bargaining. When Creditor has an exit option, Creditor's share of Firm rises dollar for dollar as the asset's liquidation value rises. Creditor receives the liquidation value, and Manager gets the entire going-concern surplus.

If Creditor actually exercised its exit option, the value of Firm as a going concern would be lost. As long as the risk of bargaining failure is small, however, Creditor's ability to lift the automatic stay or confirm a liquidating plan affects only how large a share Creditor receives in the bargain struck with Manager. This model helps us identify exactly how denying Creditor the ability to reach assets affects the outcome of the negotiations between Creditor and Manager. When Creditor has no exit option and Manager's exit option is less than the bargained-for share, Creditor's share of Firm remains constant as long as Firm has value as a going concern. No matter how much the liquidation value of the assets increases relative to the value of Firm as a going concern, Creditor continues to receive only its bargained-for share, rather than the liquidation value of the assets. As soon as the liquidation value of the assets equals the value of Firm as a going concern, however, Creditor will be able under existing law to lift the automatic stay and thus reach the collateral.[8] The payoff to Creditor suddenly jumps from its bargained-for share to the liquidation value of the assets.

It is hard to find a normative justification for a bargaining regime in which the liquidation value of Creditor's collateral is irrelevant until the extra value that Firm has as a going concern shrinks to nothing. By contrast, when Creditor does have an exit option and the liquidation value exceeds the value of Creditor's bargained-for share, Creditor's share increases as the liquidation value of the assets increases. When legal rules give parties exit options instead of imposing on them a vaguely defined duty to negotiate, the outcome is more likely to turn on things that might plausibly be part of the ex ante bargain, rather than on the relative patience of the parties, which does not seem related to any concern the parties would have at the time of their original loan.

Collective Bargaining and Exit Options

The National Labor Relations Act (NLRA), unlike its European counterparts, has not changed the substantive rights that managers and workers had under preexisting law. Rather, it aims to provide a structure in which collective bargaining can take place. Its ambitions are two-fold. First, it tries to create a bargaining environment in which strikes and labor unrest (and the costs that come with them) can be avoided. Second, and more elusively, it tries to ensure that both parties "bargain in good faith." To the extent that the law dictates bargaining rules that govern the course of subsequent renegotiations, parties may adjust other terms of their initial contract to take this into account. In the end, these rules may not benefit one side or the other, even if they work to the advantage of one side during the subsequent bargaining. In this section, however, our focus is more narrow. We look only at how legal rules can affect the dynamics of the bargaining itself.

To understand collective bargaining, we need to identify what the parties are bargaining over and what alternatives each has to reaching an agreement with the other. The simplest case to imagine is one in which the firm has no hard assets and its entire value as a going concern consists of the firm-specific skills of its workers. For their part, the workers have no alternative wage, or at least no alternative wage remotely similar to what they are making by working for the firm. All of this is common knowledge. One can imagine the stakes in this game being the discounted present value of the income stream that the firm earns over time. One can also model the bargaining that takes place as a simple Rubinstein bargaining game in which the firm and the workers' union exchange offers. We would expect that the two would reach agreement in the first period and that each would receive a share of the earnings that turned on their relative levels of patience. In the simplest case, one would expect that they would divide the revenues equally between them.

We shall focus first on the exit options that are available to the employer. The most extreme option lies in the ability of the employer to sell off the assets of the firm. *NLRB v. Burns International Security Services, Inc.*, however, held that a buyer of a firm's assets has the same duty to bargain with the union as the previous employer if the buyer "acquired substantial assets of its predecessor and continued, without interruption or substantial change, the predecessor's business operations."[9] A new buyer can put the assets to a different use or completely change the way the firm is organized. The new buyer can also hire new workers. But the buyer cannot rid itself of a union and the collective bargaining agreement if it wants to take advantage of the skills the existing employees have developed in their current jobs.

A decision that freed any new buyer from a duty to bargain with the union would change the exit option of the employer dramatically. If a new buyer can reach a much better deal with the workers, the existing employer has a threat that should improve its position in the bargaining. The question, of course, is whether this threat is credible. If the managers who bargain with the union would lose their jobs in any sale of the firm, the threat to sell the assets may not ever be carried out. As long as the union knows this, the threat would not matter much in the bargaining.

The effect of *Burns* on the dynamics of bargaining is likely to change as business conditions change. The exit option it offers the employer turns on whether the employees have firm-specific skills doing their current jobs. The more volatile the economic environment and the more a firm's operations can be restructured, the more likely it will be that a buyer will acquire the firm free of the obligation to bargain with the union, and hence the more potent the employer's threat to use this exit option. In the basic Rubinstein bargaining model, an exit option is irrelevant if it provides the player with less than the player's bargained-for share. The ability to sell assets is not likely to loom large whenever employees have substantial firm-specific skills. The greater the workers' skills, the less attractive it becomes for an employer to exit the bargaining by selling the assets. When workers have these skills, new buyers would want to keep them. The threat to sell the assets to someone who would replace the existing workers under these conditions is not credible.

When there is an economic strike—a strike over wages or other terms of employment—and the employer has not committed an unfair labor practice, the employer's most important exit option may be the ability to hire permanent replacement workers. The NLRA does prevent an employer from retaliating against its workers for going on strike, but the Supreme Court interpreted this mandate narrowly in *NLRB v. Mackay Radio and Telegraph*.[10] In this case, the Court took the view that the National Labor Relations Act does not deprive the employer of the right to "replace the striking employees with others in an effort to carry on the business"; moreover, the employer "is not bound to discharge those hired to fill the place of the strikers" after the strike is over.

A simple model shows how this exit option may affect the bargaining. Let us assume that the employees have developed firm-specific skills. The employer's ability to hire replacement workers puts a floor on how much it will receive in any negotiations. The firm is worth $1.5 million if it does not have the benefit of its current work force, but $2.5 million if it does. The employer has no ability to sell the assets to a third party. If the workers have the power to strike and shut the firm down and if the employer has no viable exit option, the outcome of the bargaining will be a division of the $2.5 million between them, with the share of each turning on their relative levels of patience. The employer and the workers will divide the firm evenly if they have equal levels of patience.[11] If the employer has the ability to hire replacement workers, it will determine the value of the firm with these workers ($1.5 million). This amount is larger than its bargained-for share if both the firm and the employees have equal levels of patience. If this is the case, the employer will receive this value and the workers will capture the difference.

Under this bargaining model, the workers capture all the surplus value they bring to the firm even in a regime in which the employer has the right to hire replacement workers. When the value of the firm without the existing employees is less than the bargained-for share (as it might well be in cases in which the value that the workers brought to the firm was unusually large), the ability to hire replacements is irrelevant because the employer's threat to replace the workers is not credible.

The Rubinstein bargaining model confirms the obvious intuition that the right to hire replacement workers improves the bargaining position of the employer relative to that of the union. The additional insight it provides lies in the way it forces us to examine the relationship between the dynamics of the bargaining and the legal rules. When a legal rule gives a party the ability to exit the bargaining process, the importance of the legal rule is determined by the value of the exit option relative to what a party would receive without it. For example, the right to hire replacement workers may not matter when the value of the firm to the employer without the existing workers is so low that the threat to hire replacements is not credible.

To justify giving or denying an employer the right to hire replacement workers, one needs to understand the bargaining process. One might, for example, want to ensure that employers could not take advantage of the workers' inability to transfer their skills to another employer. The right to hire replacement workers, however, may not enhance the employer's ability to do this. In the model we have set out, for example, the right to hire replacement workers does not itself allow the employer to capture the difference between the value of the workers' skills in this job and their value in any other.

One must, however, be even more cautious than usual in drawing conclusions from a simple model. One of the central predictions of the model is that the exit option itself does not affect the dynamics of the bargaining, but only puts a ceiling or a floor on what a player receives. This prediction is not an obvious one, and we cannot be sure that other models would generate the same prediction or that this prediction would be borne out in practice. Many have the intuition that a player should be able to use an exit option to push the bargained-for share to an amount that is even higher than the exit option itself. As with any other model, its predictions should be tested before we act on them.

A more realistic model may give the union the additional option of continuing to work at the old wage for the next period of bargaining. The ability of the workers to obtain their Rubinstein bargaining share turns crucially on the credibility of their commitment to strike. The game in which the workers have the option to work for an additional period does not have an obvious course of play. There are several subgame perfect equilibria, including one in which the workers accept a contract at the old rate and others in which there is no agreement during the first period, but in which the workers actually go on strike.[12] These models make one less certain about the nature of the division between the employer and the workers. A richer model would also take into account another kind of legal regime—one in which the firm could hire temporary workers and run its plant during a strike, but could not keep the workers after the strike was over.

Summary

In this chapter, we looked at another way in which legal rules matter, even though their direct effects cannot be seen in equilibrium. A right to specific performance in *Peevyhouse* might matter, not because the Peevyhouses would ever invoke their right to specific performance, but rather because their ability to demand specific performance affects the outcome of negotiations between the parties. Many legal rules, ranging from the perfect tender rule in contract law to the intricate rules in the Bankruptcy Code on confirming a plan of reorganization over the objection of creditors, matter not because courts are asked to apply them often, but because they establish the contours of the negotiations that take place between the parties. They and many other legal rules give one party or the other an exit option, a power to walk away from the bargaining table and still receive something by invoking a legal right. Models such as the one we developed in this chapter show how legal rules affect the bargains that people strike. In the next chapter, we explore the dynamics of bargaining further and focus in particular on the way in which nonverifiable information must be taken into account.

Bibliographic Notes

Game theory and bargaining. There is a vast literature on bargaining theory in economics, only a small fraction of which is touched upon in this book. A radically different alternative to the Rubinstein alternating offers model is known as the *axiomatic approach*. Instead of looking at the dynamics of the bargaining process between two self-interested parties, it begins by positing the characteristics that an agreement between the parties should have. The next step is to examine all the agreements that are possible in a particular case. (In our example with the book, the possible agreements are all the sale prices from $10 to $15.) One also notes the consequences if no agreement is reached. (The seller keeps a book that the seller values at only $10 and the buyer keeps cash that would otherwise have gone to the sale.) The next step is to identify outcomes that are consistent with the characteristics we determined an agreement should have.

John Nash, whose equilibrium solution concept has been a recurring theme in the first six chapters, was a pioneer in this area as well. He set out a series of characteristics, or *axioms*, all of which seem plausible, and then showed that, in any bargaining situation, only one outcome could satisfy all of his axioms; see Nash (1950a). This way of examining bargaining is known as *cooperative bargaining* and has recently attracted growing attention. One of the great challenges in modern game theory is carrying out what is called the *Nash program*, which is to connect the principles of cooperative bargaining with those of noncooperative bargaining.

Rubinstein (1982) is the origin of the basic Rubinstein bargaining model. As a general guide, Osborne and Rubinstein (1990) is a good rigorous examination of the foundations of the basic bargaining model of this chapter, and it provides many extensions and elaborations thereon. A clear exposition and easily accessible proof is in Kreps (1990b).

There is a vast experimental literature on bargaining which indicates that negotiating behavior does not always conform to the economic model. Irrational behavior, social norms, and cognition problems have been documented. Good places to become familiar with this work are Raiffa (1982) and Neale and Bazerman (1991).

Legal rules as exit options. Sutton (1986) modifies the Rubinstein model by using exit options. Baird and Picker (1991) shows how legal rules can be modeled as exit options. *Peevyhouse* is discussed in Posner (1992). Cooter, Marks, and Mnookin (1982) models bargaining in the shadow of legal rules by positing that those engaged in bargaining have different characteristics, only some of which are observable.

Private information and exit options. The field of literature on private and asymmetric information is large. For variations on the Rubinstein model, see Rubinstein (1985), or Ausubel and Deneckere (1989) on bargaining with incomplete information. Many of the results in this chapter rely upon the Rubinstein bargaining model. Other potential lines of research into bargaining with private information would include *cheap talk* (statements by parties that are costless, nonverifiable, and nonbinding, but that nevertheless convey information) and other refinements. See Farrell and Gibbons (1989), Palfrey and Rosenthal (1991), Rabin (1990), Seidmann (1990), or Matsui (1991). In addition, there is an entire literature on optimal mechanism design that tackles issues similar to those involved in designing a legal system around an information structure; see, for example, Myerson (1979) or Myerson (1981).

Bargaining and corporate reorganizations. For a discussion of the general bargaining problem that arises in the context of bankruptcy, see Mnookin and Wilson (1989). Baird and Picker (1991) incorporates exit options into the Rubinstein bargaining model to examine the law of corporate reorganizations. Bergman and Callen (1991) and Bebchuk and Chang (1992) use the Rubinstein model to examine corporate reorganizations, but they do not use exit options. Bergström, Högfeldt, and Lithell (1993) examines bargaining among creditors and a common debtor using a Nash bargaining model.

Collective bargaining and exit options. The literature using strategic bargaining models is large and growing. We make specific use of Fernandez and Glazer (1991).

CHAPTER 8

Bargaining and Information

Basic Models of the Litigation Process

In this chapter, we look at how legal rules alter the dynamics of bargaining when one player possesses information that the other does not. The existence of pooling or separating equilibria determines not only what each party receives in the end, but also whether parties reach agreement at all. In this chapter, we use these rules of civil procedure as our primary focus in exploring bargaining between the parties when there is private, nonverifiable information.

The litigation of a dispute under Anglo-American law follows a definite and well-worn path. Although the details differ slightly between the state and federal court systems and from one state to another, the basic elements are the same. In a simple case in which Victim wants to recover damages from Injurer, the lawsuit begins when Victim files a complaint in which Victim provides "a short and plain statement" of why Victim is entitled to damages from Injurer.[1] Injurer is required to file an answer that states "in short and plain terms the party's defenses to each claim asserted and shall admit or deny the averments upon which the adverse party relies."

Next, the parties engage in *discovery.* After a set of initial disclosures that each party may be required to give the other, each party may request information from the other. The "request shall set forth the items to be inspected either by individual item or category, and describe each item and category with reasonable particularity."[2] At any point, one or both parties may assert that the evidence, construed in a light most favorable to the other, is such that the party is entitled to prevail and move for summary judgment.[3]

If these motions fail, the case proceeds to trial. The judge decides questions of law. Either the judge or a jury decides questions of fact. In most cases, questions of liability and damages are heard simultaneously. In some cases, however, the jury decides only whether a party is liable. If the defendant is held liable, there is a separate trial on the question of damages.[4] After the trial, either party may file an appeal, which is typically limited to whether the lower court correctly applied the law to the case and whether there was a substantial basis in the evidence presented to support the decision of the fact finder. If Victim prevails at trial, Victim is entitled to have a judgment entered.

Once judgment is entered, Victim is entitled to call upon the state to seize property of Injurer (typically by having the clerk of the court issue a writ of execution to the sheriff) if Injurer does not pay the damages voluntarily. At any point during the process, and indeed before the formal suit is filed, the parties may settle the dispute on any terms they find mutually acceptable. In this country, parties typically must pay their lawyers out of their own pockets. This rule is frequently contrasted with the so-called *English rule,* in which the losing party must pay the fees of the other.

The overwhelming majority of civil cases settle before trial. Many settle before they are filed. The rights that parties have at trial, however, determine the terms of settlement. Moreover, the rules determine how and when parties incur costs during the course of litigation. A change in the rules may alter not only the costs that each party bears but also when during the course of the litigation each cost must be incurred. Both kinds of changes can have powerful effects on the dynamics of litigation. Finally, the rules of procedure are structured in such a way that parties learn more information as the case proceeds. The way in which parties acquire information and the costs they incur in gathering information also have a powerful effect on the dynamics of litigation.

We first review the two basic ways in which civil litigation has been modeled, the *optimism model* and the *private information model.* In the section that follows, we show how these models shed light on the question of how breaking down litigation into smaller components, such as separate hearings on liability and damages, affects the behavior of parties in bringing lawsuits and in settling them. Our model highlights a theme that we have already encountered a number of times: Altering the sequence or the number of moves in a game has a dramatic effect on the likely course of play. Hence, the effects of even small changes in the rules of civil procedure are often significant. The next section shows how private information also affects the kinds of cases that are likely to be litigated. In the final section, we examine rules governing discovery and return to the problem of information, particularly, the differences between verifiable and nonverifiable information.

Litigation is costly. In addition to the direct costs of court and legal fees, litigation consumes the time and energy of the litigants. One might therefore predict that rational litigants would find it in their interest to settle, based on the award the plaintiff could expect to win in the event of litigation, adjusted by the likelihood that the plaintiff would succeed. The parties could then divide the savings in litigation costs between them. Consider a typical tort suit. Injurer's car ran over Victim and Victim suffered $100,000 in damages. Victim exercised due care and Injurer did not. Litigation would cost each party $10,000. If the parties were to negotiate with each other at the outset, they should reach some agreement. Injurer should prefer settling rather than litigating at any amount less than $110,000, and Victim should prefer settling rather than litigating at any amount greater than $90,000. There is $20,000 of surplus that the parties could split if they avoided litigation. Although each party would want to appropriate as large a share of the surplus as possible, it seems likely—and the analysis in Chapter 7 suggests—that they would agree on some division.

Most litigants do, in fact, settle. We want to focus, however, on the minority of cases in which parties do litigate and ask whether changes in the legal rules would affect the likelihood of settlement. We need to relax the assumptions that we have made. The only conclusion we might draw so far is that the size of the surplus to be divided between the litigants affects the likelihood of settlement. It is possible, for example, that agreement will be more likely as legal fees increase. In other words, if two parties are bargaining over a $10,000 bill that is sitting on a table and neither gets it if they fail to agree, they are more likely to reach agreement than if there is a $1 bill on the table. The greater the potential gains from trade, the harder each party will work to ensure that there is an agreement. This intuition, however, is not self-evident. The benefits from posturing and acting tough—and expending resources to appear that way—are much greater if there is $10,000 on the table than if there is $1 on the table.

We need to ask which of the assumptions that we have made need to be relaxed. Introducing uncertainty does not itself change matters. Return to our example in which Victim suffered $100,000 in damages, but assume that the parties are not certain that Injurer is liable. If the probability of Victim's prevailing in court is only 75 percent, the only thing that changes is the amount of the settlement—now somewhere between $65,000 and $85,000.[5]

We can, however, explain why some cases do not settle, if we assume that there is not just uncertainty about the outcome of litigation, but also disagreement about the likely outcome. For example, both sides may agree that damages are $100,000, but Victim may think that the probability of prevailing is 90 percent while Injurer thinks that the probability is only 60 percent. Victim expects to have $80,000 at the end of the litigation, but Injurer expects to pay only $70,000.[6] There is no offer Injurer is willing to make that Victim is willing to accept. Litigation ensues because the settlement range has disappeared. Models of litigation built on the idea that parties have different views about the likely outcome are known as optimism models because the failure to settle arises from one or both sides in a particular case being overly optimistic about the outcome of litigation.[7]

Instead of one or both parties being too optimistic, however, it is also possible that parties fail to settle because one has information that the other does not. The optimism model relies on differences in opinion, not differences in information. The distinction is subtle but fundamental. A party's beliefs are unaffected by another party's opinion, but they are affected by another party's information.

Let us go back to the numerical example we gave above. Assume that Injurer believes that there is only a 60 percent probability of losing because of private information. Injurer, for example, may have a better sense of whether a court or a jury can be persuaded that Injurer failed to exercise due care. Injurer may also have studied similar cases and have a better sense than Victim of how a court is likely to rule. This difference is one in information rather than belief because, if there is a way for the defendant to convey this information credibly, Victim's beliefs will change—as will Victim's expectations about the outcome of litigation.

The distinction between differences in information and opinion affects the way we think about pretrial bargaining and litigation. When different beliefs are information-based, the rules governing pretrial negotiation may determine whether information is conveyed from one party to another. In our example, Injurer has two reasons to convey the information to Victim. First, once Victim's estimate of damages comes close enough to Injurer's, a settlement range will exist where none existed before. Even if the parties would have settled in any event, Injurer should want to convey the information for a second reason.

Victim is likely to accept a smaller settlement once Injurer can convince Victim that the likelihood of success is only 60 percent.

Unlike models based on the opinions of the players, information models allow one to analyze incentives for information revelation and the effects of different disclosure and discovery rules. Later in this chapter, we develop a model to distinguish the effects of mandatory disclosure and other discovery rules. In some situations, however, Injurer's information will be nonverifiable. In this case, Injurer cannot communicate the information directly, but information may be signaled during the course of pretrial negotiations. Injurer's willingness to make a settlement offer or reject one may signal information.[8]

To continue with our example, assume that Injurer is one of two types, depending on how much care was exercised at the time of the accident. The first type of injurer we shall call "negligent." This injurer exercises so little care that a jury finds in favor of Victim 80 percent of the time. The second type we shall call "careful." This injurer is sufficiently careful that a jury is likely to find in Victim's favor only 20 percent of the time. Assume further that Injurer's type is not apparent to Victim and becomes evident only after Injurer takes the stand in open court.

Victim does not know which type of injurer was involved in the accident, only that there are equal numbers of each type in the population. Hence, before any negotiations, Victim thinks that the likelihood of winning is 50 percent. We continue to assume that both players know that Injurer will owe Victim $100,000 if the jury finds Injurer liable and that each side must pay $10,000 in legal fees if the case goes to trial.

We can illustrate the dynamics of pretrial negotiations by examining two different extensive forms for the game. First, assume that Injurer can make a take-it-or-leave-it offer. If Victim rejects the offer, they go to trial. The appropriate solution concept is a perfect Bayesian equilibrium because this is a dynamic game of incomplete information. We restrict attention to pure strategies, looking for a separating equilibrium in which the negligent injurer makes an offer that is different from the one made by the careful injurer.

In any separating equilibrium, only one type of injurer can settle. We can see this by positing that an equilibrium exists in which the negligent and careful injurers make different offers, but Victim accepts both. In that event, the negligent injurer who is making a higher offer is not making a best response. In this proposed equilibrium, Victim believes that anyone who makes the low offer is careful. A negligent injurer could mimic the careful injurer's lower offer and such an offer would be accepted. Hence, a strategy combination in which the negligent and the careful injurers make different offers and both are accepted cannot be a solution to the game.

If a separating equilibrium exists in this game in which Injurer makes a take-it-or-leave-it offer, it must be one in which only the offer of the negligent injurer is accepted. In such an equilibrium, Victim would accept an offer from a negligent injurer only if the injurer's offer exceeded what Victim expected to receive in the event of litigation. Under our assumptions, this amount is $70,000. (Victim has an 80 percent chance of winning $100,000 from a negligent injurer but will spend $10,000 on litigation, giving a net return of $80,000 − $10,000 = $70,000.) By contrast, a careful injurer would be willing to offer no more than $30,000. (The careful injurer has a 20 percent chance of having to pay $100,000 to Victim. In the event of litigation, however, the careful injurer will have to spend $10,000, regardless of the outcome. Hence, the careful injurer's expected loss from going to trial is $20,000 + $10,000 = $30,000.) Thus, there is a separating equilibrium in which the negligent injurer offers $70,000 and the careful injurer makes no offer or one that is not accepted. Victim accepts all offers greater than or equal to $70,000.

We must now look for pooling equilibria. In this example, there is no pooling equilibrium in which both types of injurer make the same offer and it is accepted. Victim would accept a pooling offer only if it were greater than the amount Victim could expect if no one settled and there was litigation in all cases. This amount is $40,000.[9] The careful injurer, however, is unwilling to make any offer greater than $30,000. Because only the negligent injurers will be willing to pay $40,000, there is no pooling equilibrium in which both injurers make an offer that Victim accepts.

Second, assume that Victim can make a take-it-or-leave-it demand. There are three possibilities. Victim can make a demand that both types of injurer accept, that only one type accepts, or that neither accepts. The negligent injurer is the one who accepts when only one type accepts because the negligent injurer is the one who expects to do worse if there is a trial and therefore the one who finds settlement more attractive.

The negligent injurer will accept any demand less than $90,000 and the careful injurer will accept any demand less than $30,000.[10] If Victim demands $30,000, both types of injurers accept the offer and Victim receives $30,000 with certainty. If Victim demands $90,000, the negligent injurer accepts it, but the careful injurer rejects it. (The careful injurer rejects because the offer exceeds the $30,000 the careful injurer expects to lose in the event of litigation.) Victim will prefer to make the separating offer of $90,000 to the pooling offer of $30,000. The expected return to Victim on the pooling offer is $30,000, whereas the expected return to Victim on the separating offer is $50,000. (In half the cases, Victim settles for $90,000 and spends no money on legal fees. In the other half, Victim goes to trial, spends $10,000, and expects to win 1 time in 5. Hence, the expected return to Victim from offering $90,000 net of litigation costs is $50,000.)[11]

The expected return from the separating offer is also greater than the $40,000 Victim would receive from making an offer no one would accept and going to trial in all cases. The solution to the game, therefore, is the separating equilibrium in which Victim makes the $90,000 demand. The negligent injurer accepts it, and the careful one litigates. Victim's expected gain is $50,000, while the negligent injurer's expected loss is $90,000, and the careful injurer's expected loss, including litigation costs, is $30,000.

We have examined two radically different extensive form games. Whether the injurer or the victim has the ability to make take-it-or-leave-it offers affects the payoffs dramatically, but the equilibrium in both games has the same structure. In both games, the negligent injurer settles and the careful injurer litigates. The only difference is that the gains from settling rather than litigating shift from one player to the other. (The opportunity to make a take-it-or-leave-it offer creates bargaining power, which in turn allows the offeror to extract the gains from trade. These are the $20,000 savings in litigation costs that settlement brings.) These models focus not so much on how the gains from trade are divided as on how private information itself affects the dynamics of bargaining. They suggest that the defendants in litigation who are least likely to be found liable spend the most in litigating the cases. Only by being willing to spend resources on litigation can they signal their type.

It is worth comparing bargaining in the shadow of a legal rule with other kinds of bargaining. Consider, for example, a simple model of Seller and Buyer. Buyer is one of two types (one with a high or a low reservation price). The solution to the game between Buyer and Seller turns on whether Seller or Buyer can make a take-it-or-leave-it offer. If Buyer (the player who possesses private information) makes the offer, Buyer will offer Seller an amount equal to Seller's reservation value (which is common knowledge) and thereby capture all the gains from trade. This offer is always accepted. When Seller—the uninformed player—has the ability to make the take-it-or-leave-it offer, Seller typically makes a screening offer that allows some bargaining failure, but extracts more surplus than if an offer were made that would always be accepted.

The game between Victim and Injurer has a similar structure, but the solution—in which some bargaining failure takes place no matter who makes the offer—is not the same. The situation of the uninformed player in the event that no bargain is reached is different. When no bargain is reached, Seller enjoys no gains from trade. Seller is left in the same position, regardless of whether Seller is dealing with a buyer who has a low reservation price or a high one. When the sale does not take place, the different reservation prices of the two types of buyers do not affect the payoff Seller receives. By contrast, Victim's payoffs when no bargain is reached depend on the type of injurer Victim faces. Litigation follows when there is no settlement and the outcome of litigation for both Victim and Injurer turns on the private information that Injurer possesses. The effect of private information on both parties in the event that no bargain is reached explains why the solution to this game is different from the one involving Buyer and Seller.

Modeling Separate Trials for Liability and Damages

In this section, we examine the effects of splitting a trial into separate components and show how information problems affect pretrial bargaining.[12] We begin with a model in which parties have different opinions rather than different information. Let us assume that Victim thinks that the chances are 9 in 10 that a jury will find Injurer liable and that, if the jury does find Injurer liable, the damage award will be $110,000. Injurer thinks that there is only an 80 percent chance of being found liable and that, if found liable, the damages will be only $100,000. In a sequential trial regime, the parties can settle entirely, stipulate damages and litigate liability, stipulate liability and litigate damages, litigate both, settle damages after litigating liability, or settle liability after litigating damages. In a unitary trial regime, the last two options are not available. We assume that the costs of each stage of a sequential trial are $5,000 for each side, and the costs of a unitary trial are $10,000 for each side.

In a legal regime in which there are unitary trials, the parties are willing to settle the entire lawsuit. Victim's expected recovery of $89,000 is less than Injurer's expected costs of $90,000.[13] The settlement range in this example, however, disappears in a legal regime in which liability is tried first and then damages. If we assume that the costs of litigation are the same in a case that is fully tried, the expected costs of litigation for both litigants is lower in a regime with sequential trials. In a unitary trial regime, the litigants must spend resources introducing evidence about damages in all cases; in a sequential trial regime, they must spend these resources only if the defendant is found liable. The less that parties can expect to save on litigation, the more their expectations about the outcome of any litigation must differ in order for settlement to occur. In our case, for example, Victim's expected recovery net of cost ($94,500) is higher and Injurer's expected cost ($84,000) is lower in a sequential rather than a unitary trial regime.[14] Because of their different beliefs about Victim's chances, they both prefer to litigate rather than to settle.

Sequential trials also produce another effect. Because the expected expenses of litigation are lower, some suits that might not be worth bringing under a unitary trial regime would be worth bringing under a regime in which liability and damages were tried separately. Assume, for example, that Victim thinks that there is a 90 percent chance of winning $11,000. A unitary trial again costs $10,000, and each phase of a sequential trial costs $5,000. In a unitary trial regime, the case will not be brought because it has a negative expected value.[15] It would be brought in a sequential trial regime, however, because there it has an expected value of $400.[16] The costs are $500 lower in a sequential trial regime because the plaintiff avoids spending $5,000 litigating damages in the 10 percent of the cases in which the plaintiff loses in the liability phase.

These models suggest that the choice between unitary and sequential trials turns on the relative size of the different effects. The overall effects on welfare are ambiguous. They depend first on how much costs increase. Costs increase for two reasons. First, there is an increase in the number of cases that are filed in a regime of sequential trials. Second, there is a reduction in the number of cases settled, the exact number turning on how often a settlement range would exist under a unitary trial regime, but not under one with sequential trials. Against these costs, however, we have to balance the savings that arise when cases are litigated, the injurer is found not to be liable, and the expenses of litigating damages are saved.

At this point we turn from the optimism model to one that focuses on information. We need to take additional considerations into account if parties have different views about the victim's prospects because each has information the other does not. Assume again that there are two parties: Victim and Injurer. There are two types of injurers and two types of victims. Injurer is either negligent and knows there is a 75 percent chance of being found liable or Injurer is careful and knows there is only a 25 percent chance of being found liable. Victim believes that Injurer is as likely to be one type as another. Victim is also one of two types. One type of victim suffered damages of $100,000 and knows it. The other type suffered damages of only $60,000 and also knows this. Injurer knows only that Victim's damages are either high or low with equal probability. Litigation costs are $5,000 per side for each stage of the trial. Each side pays $5,000 in legal fees if only liability is litigated, but pays $10,000 if there is a unitary trial for both liability and damages or if there are separate trials for both.

We also adopt a simple extensive form for the pretrial negotiations. Injurer can make a take-it-or-leave-it settlement offer to Victim. Victim can accept or reject the offer. If Victim accepts, the terms are followed and the game ends. If Victim rejects, they proceed to trial. In the unitary trial regime, both liability and damages are litigated. In the sequential trial regime, liability is litigated. If Victim prevails with respect to liability, Injurer can make another take-it-or-leave-it offer prior to the damage trial. If the offer is rejected, Victim and Injurer litigate damages. In a regime of a single trial for liability and damages, there is only one settlement offer. In the sequential trial regime, there can be two different settlement offers: one before the liability trial and one before the damages trial.[17]

We begin with a unitary trial. We must look for both separating and pooling equilibria. If there is an equilibrium in which the two types of injurers adopt different strategies, the negligent injurer makes an offer that either both types of victims accept or just the low-damage victim accepts. In either case, the victim's beliefs must be that the offer comes from a negligent injurer, given that we are looking for a separating equilibrium. If the negligent injurer makes an offer that both types of victims will accept, it must be at least $65,000, the minimum offer the high-damage victim will accept from a negligent injurer. If a negligent injurer makes an offer that only the low-damage victim will accept, it must be at least $35,000.[18]

Let us compare the expected payoffs to the negligent injurer from the two offers. Both types of victims accept the first offer of $65,000. By contrast, only ½ of the victims (those with low damages) accept the second offer of $35,000. To decide whether the lower offer is better for the negligent injurer than the higher one, we must take into account the amount that the negligent injurer can expect to lose when high-damage victims reject the offer and then force the negligent injurer to litigate against them. When this amount ($85,000) is considered, the negligent injurer's expected costs from making the settlement offer of $35,000 are only $60,000.[19] Hence, the negligent injurer prefers making this offer, one that only low-damage victims accept, rather than making one that they both accept.

An equilibrium that separates the different kinds of injurers can exist only if the careful injurer prefers litigating to mimicking the negligent injurer's strategy (making an offer that low-damage victims accept and high-damage victims reject). Such a strategy brings an expected cost of $35,000. By litigating each case to the end, however, the careful injurer expects to lose only $30,000.[20] Hence, the careful injurer would rather litigate than follow the lead of the negligent injurer. A separating equilibrium therefore exists in which the negligent injurer and the low-damage victim settle, and everyone else litigates.

We now must check for a pooling equilibrium in which both types of injurers make the same offer. We can eliminate the possibility that both injurers can make an offer that both types of victims will accept. Both types of victims accept if the high-damage victim accepts. The high-damage victim, however, accepts only if the offer exceeds the expected recovery against a pool that consists of equal numbers of negligent and careful injurers. This amount, taking litigation costs into account, is $40,000.[21] The careful injurer, however, expects to pay only $30,000 in the event of litigation. Hence, the careful injurer will never be willing to join the negligent injurer in making an offer that the high-damage victim accepts.

There is, however, a pooling equilibrium in which both types of injurers make an offer that only one type of victim accepts. The low-damage victim accepts an offer from both types of injurers if it exceeds the expected net recovery in the event of litigation ($20,000). Both types of injurers are willing to make such an offer. If the negligent injurer makes this offer and it is accepted, the negligent injurer faces expected costs of only $52,500. This amount is far less than the costs of $70,000 that the negligent injurer can expect to pay in the event that both types of victims litigate.[22] The careful injurer is willing to make such an offer as well. The careful injurer could litigate against both types of victims and expect to lose $30,000, but the careful injurer is better off settling with low-damage victims for $20,000 and then litigating against high-damage victims. In this case, the careful injurer's expected losses are only $27,500.[23]

In this model, there are two equilibria—one a separating equilibrium in which negligent injurers settle with low-damage victims and in which careful injurers make no settlement offer at all, and the other a pooling one in which both types of injurers make an offer that only low-damage victims accept. The first solution to the game, however, does not seem plausible. We can see this by noting first that each type of injurer pays less in the pooling equilibrium. Bearing this in mind, consider the following deviation in the separating equilibrium: Both types of injurers deviate to the pooling equilibrium offer. In the proposed separating equilibrium, Victim believes at the outset that both types exist in equal numbers. Victim should not change this prior belief about those who deviate off the equilibrium path because both types of injurers strictly prefer the pooling offer if the low-damage victim accepts, and a low-damage victim who has these prior beliefs will accept. Because Victim would not change this belief in the face of the deviation, neither player chooses a best response in this proposed equilibrium. Each player can enjoy a higher payoff by choosing another strategy.

There is another way of understanding the intuition at work. Victim should believe an injurer who makes a pooling offer and makes the following statement: "You should believe that I am equally likely to be either type because both types would strictly prefer this offer to the equilibrium if your beliefs are that we are equally likely to be either type." Because there is a consistent deviation from the separating equilibrium, it is not part of the likely course of play.

Thus, the pooling equilibrium is the solution to this game. Both the negligent injurer and the careful injurer make settlement offers of $20,000. Low-damage victims accept such offers and receive $20,000. High-damage victims have a 50 percent chance of winning against a pool of injurers that consists of equal numbers of careful and negligent injurers. The high-damage victim, however, must pay $10,000 in legal fees. Hence, the high-damage victim has an expected net recovery of $40,000. The careful injurer expects to pay $27,500, and the negligent injurer expects to pay $52,500. Because ½ of all cases settle (those involving low-damage victims), the expected litigation costs are $10,000. (Injurer and Victim spend $10,000 each in ½ the cases.)

We are now ready to analyze the sequential trial regime. As usual, we must work our way from the end of the game to the beginning. We analyze the possible positions in which Injurer might be after being found liable. The remaining issue is the amount of damages Injurer must pay in the event of litigation. At this stage, Injurer's type is irrelevant because liability has already been established. We have to take into account the three possibilities. Injurer may face only high-damage victims, only low-damage victims, or an equal mixture of both. In the first two cases, there is no private information and the parties will settle. Injurer will make an offer equal to Victim's expected recovery, less Victim's costs of litigating damages. Hence, in the case of a high-damage victim, Injurer will offer $95,000; in the case of a low-damage victim, Injurer will offer $55,000. Both offers will be accepted.

The more difficult case arises when Injurer does not know Victim's type. Both types accept offers of $95,000, but only the low-damage victim accepts an offer of $55,000. If Injurer offers $95,000, Injurer's expected costs are also $95,000 because everyone accepts and there is no litigation. If Injurer offers $55,000, however, only $\frac{1}{2}$ the victims accept (the ones with low damages) and the other $\frac{1}{2}$ litigate. Injurer's expected costs are $80,000.[24] Injurer is therefore better off making the $55,000 offer. Hence, $\frac{1}{2}$ the cases that proceed to a trial on liability and in which Injurer is found liable are settled before the trial on damages.

Now that we know the possible courses of play after Injurer is found liable, we can look at the bargaining that takes place at the outset, before there is litigation about either liability or damages. First, we look for an equilibrium in which the two types of injurers make different offers. Imagine that the negligent injurer makes an offer that both types of victims accept. In this proposed equilibrium, the negligent injurer never goes to trial on the issue of liability and never has to worry about bargaining after the liability phase.

For an equilibrium to exist in which the negligent injurer never goes to trial, the negligent injurer must believe that those who litigate are likely to be high-damage victims. In such an equilibrium, no victim would go to trial, hence litigation would be off the equilibrium path and Bayes's rules would not constrain the negligent injurer's beliefs for out-of-equilibrium actions. Note, however, that, if the negligent injurer believed that those who litigated were low-damage, the negligent injurer would not be playing a best response by making an offer that is high enough that both types accept it.

In any such pooling equilibrium the low-damage victim must receive more from settling than from litigating. Therefore, any settlement offer that the low-damage victim accepts dominates what the low-damage victim receives from litigating. Because rational players do not believe that other players adopt dominated strategies, Injurer must believe that any victim who deviates from a pooling equilibrium is high-damage until the settlement offer is greater than what the high-damage victims would receive from litigating. (At this point, this refinement no longer constrains Injurer's beliefs because a deviation on the part of either player would be an adoption of a dominated strategy.) The only settlement offer from a negligent injurer that both types of victims accept is one that gives a high-damage victim more than the high-damage victim would receive from deviating from the proposed equilibrium and going to trial. This amount is $66,250.[25]

Let us consider an equilibrium in which the negligent injurer makes an offer that only low-damage victims accept. If Injurer makes such an offer, Injurer should believe that any victim who rejects the offer has suffered high damages. An equilibrium in which low-damage victims settle must be one in which the low-damage victim's best response is to settle, notwithstanding Injurer's belief that those who litigate are high-damage. When Injurer believes that those who litigate are high-damage, however, low-damage victims would receive the same settlement offer as high-damage victims after the liability phase. Low-damage victims have no incentive to treat a settlement offer before the liability phase differently from the way high-damage victims do. Both receive the same payoff from litigating liability in this proposed equilibrium because Injurer believes that those who litigate liability are high-damage. The only offer that the low-damage victim accepts from a negligent injurer is therefore one that a high-damage victim would accept as well. If the low-damage victim were willing to accept an offer that the high-damage victim would not, the low-damage victim's strategy would not be a best response given the beliefs of Injurer. For this reason, we can reject an equilibrium in which the low- and the high-damage victims separate themselves in the liability phase of the trial.

The inability of the negligent injurer to make a settlement offer before the liability phase that separates high- and low-damage victims highlights a feature of the information model of sequential trials that distinguishes it from the optimism model. An equilibrium in which low-damage victims settle early is necessarily one in which Injurer infers that the victim who does not settle is high-damage. The ability of Injurer to draw such an inference prevents an equilibrium from existing in which the different types of victims respond to settlement offers differently. The inferences that can be drawn from the willingness to settle serve to make settlement less likely. By contrast, in the optimism model, the potential to save litigation costs drives the willingness of parties to settle.

The negligent injurer must choose between making an offer that all the victims accept or litigating with all of them. The costs of litigating liability against both types of injurers is $65,000.[26] This amount is less than $66,250, the minimum settlement offer that either victim will accept from a negligent injurer. Hence, the negligent injurer always litigates. As noted, we cannot have a separating equilibrium in which the careful injurer settles with some or all of the victims and the negligent one does not. In such an equilibrium, the negligent injurer would always want to mimic the careful one and would make the same settlement offer. There is no equilibrium in which the negligent injurer makes an offer that some victims accept that is different from that of the careful injurer.

We now ask if there is an offer that both kinds of injurers can make that induces settlement in the first stage. We know that they cannot make an offer that only low-damage victims accept. Low-damage victims will always deviate in a proposed equilibrium in which Injurer believes at the liability phase that all the low-damage victims have settled and only high-damage victims remain. We need to make only two additional calculations to show that we cannot have an equilibrium in which both types of injurers make an offer that both types of victims accept before the trial on liability. Both types of victims will accept an offer that both injurers make only if the offer gives high-damage victims what they would receive if they chose to litigate liability. This offer must be at least $42,500.[27] The careful injurer, however, would never make such an offer because the careful injurer is better off litigating liability. The expected costs of litigating liability against both kinds of victims to the careful injurer are only $25,000.[28] Thus, there is no pooling equilibrium in which both injurers are willing to make an offer that both types of victims accept.

We have shown that there is no equilibrium in which any victim settles before the trial on liability. The only equilibrium in this game arises when neither injurer makes an offer in the first stage—or makes an offer so low that no victim accepts it. After liability is determined, Injurer knows that Victim is as likely to have suffered high damages as low damages. Hence, when Injurer loses the liability phase, Injurer offers Victim $55,000. Victims who have suffered low damages accept the offer and those with high damages litigate. Low-damage victims enjoy an expected net recovery of $22,500; high-damage victims enjoy $42,500; the negligent injurer loses $65,000; and the careful injurer loses $25,000.[29] Victim and Injurer each incur $5,000 in litigation costs during the liability phase. Because only ½ the injurers are found liable and because victims with low damages settle after Injurer is found liable, there is a trial on damages in only ¼ of the cases. When it takes place,

Victim and Injurer again spend $5,000 each. Therefore, the expected litigation costs under the sequential trial regime are $12,500.

We can now look at the differences between the unitary and the sequential trial regimes. In this example, the added costs from litigating liability in every case are greater than the savings from avoiding litigating damages in cases in which Injurer is not found liable. The difference in litigation costs, however, is not a general one. Cases can also arise in which both types of victim settle in the first round. The persistent difference between a unitary and a sequential regime lies in the way the prospect of subsequent rounds of bargaining induces parties to conceal private information. When the players possess private information, they are unlikely to signal it if the game has multiple stages and the private information will affect the course of bargaining in subsequent rounds.

A low-damage victim has an incentive to mimic the high-damage victim, because high-damage victims receive large settlement offers after they prevail during the liability trial. The low-damage victim, in other words, hides information that would be harmful in subsequent bargaining. A low-damage victim will not accept a low offer if rejecting that offer induces the belief that the damages that the victim has suffered are high. Parties do not have to worry about subsequent rounds of bargaining in a unitary trial regime. Hence, they have no reason to reject or accept offers in order to hide private information. The failure to settle before the start of litigation may signal to Injurer that Victim has suffered high damages, but this information does not affect Injurer's conduct because the time for settlement has passed. In a unitary trial regime, the low-damage victim has no incentive to mimic the behavior of the high-damage victim because it does no good. At this point in a unitary trial, Injurer proceeds against Victim the same way, regardless of type.

This example illustrates a basic force that is at work in pretrial bargaining. If we believe that one of the goals of a civil procedure regime is to promote the transmittal of information and enhance settlement at values related to the strength of the case, there appears to be an argument for combining rather than splitting trials. The more phases of litigation (including pretrial phases), the less incentive a party has to signal private information. Very casual empiricism suggests that much litigation does not settle until after large costs have been incurred. One of the most expensive aspects of litigation is discovery, and many lawsuits settle only after discovery is largely completed. In similar fashion, bankruptcy reorganizations take a great deal of time and rarely settle before a variety of issues, such as priority and validity of claims, are litigated.

Information and Selection Bias

Legal analysts commonly draw inferences about legal rules from a study of litigated cases. In order to draw the correct inferences from such a study, however, it is important to know the extent to which the few cases that are litigated differ systematically from all the disputes that arise. If these suits are not representative of the population of disputes, as they almost certainly are not, one must take account of the selection effect, which might also influence the way judges decide cases. Rules that are optimal for the class of parties that come to court might make little sense if one took into account the effect of these legal rules on parties who settled. Because these cases by definition never reach the courts, judges will not be able to observe directly how the rule affects these litigants. Thus, it is important to understand how this selection effect operates in order to control for biases in litigated disputes.

The optimism model and the information model have different implications for the types of cases that are settled rather than brought to trial. In any model in which the differences in expected outcomes are based on differences of opinion, a dispute will be litigated whenever the plaintiff is sufficiently more optimistic than the defendant about the outcome. There will be a selection bias if certain types of disputes are more likely to result in large differences in opinion about trial outcomes. For example, the cases in which differences in opinion are large may be cases in which each party is equally likely to win.[30]

We would not, however, expect the same outcome if the bargaining process induced parties with certain types of information to signal it. Consider again the case in which the injurer has better information about liability than the victim. We saw that, in equilibrium, the negligent injurer settles and the careful injurer litigates. The implication for selection bias is very different from that of the optimism model. There is a systematic shift toward less success at trial for the victim than in the underlying population of disputes. Those injurers most likely to be found liable have already settled; those least likely to be found liable cannot signal their type and hence must incur the expense of a trial. This shift is independent of whether the underlying population consists of disputes that are likely to result in liability or unlikely to result in liability.

Other information structures can give different implications. Consider cases in which the injurer has better information about liability and the victim has better information about damages. In pretrial bargaining the negligent injurer will be more willing to make an offer that the victim will accept (or accept a given offer from the victim). Similarly, a victim with low damages is more willing to make an offer that the injurer will accept (or accept a given offer from the injurer). Here the selection effect is to have litigation where the probability of liability is low but the damages conditional on liability are high, relative to the underlying population of disputes.

Thus, the kinds of cases that are litigated may differ dramatically depending on the forces that determine whether litigants settle. For this reason, it is hard to adjust litigation data to account for selection bias. Any empirical data must be adjusted, and the adjustment requires some model of the pretrial negotiation process.

Discovery Rules and Verifiable Information

Up to this point, we have assumed that the private information that the litigants have is not verifiable. In many cases, however, the information is verifiable. The outcome of a trial, after all, turns on the evidence that the litigants can bring before the court. If a party believes that the prospects of prevailing are good, that party must be making some assessment about how the court and the jury will assess the information with which they will be presented. The relevant private information about a victim's damages as far as settlement is concerned is not how much the victim actually suffered, but rather how much the court and the jury will find the victim suffered. Because they make this determination on the basis of the evidence the parties present, the parties should in principle be able to present it to each other.

Of course, some information—such as a party's knowledge about how effective various witnesses are likely to be—may not be easy to reveal before trial. Litigation models in which parties have nonverifiable private information therefore serve a useful purpose. Nevertheless, litigation models should confront problems of verifiable information as well as private, nonverifiable information. Written documents make up much of the evidence that is put before a court in the typical commercial law dispute. Such documents are verifiable information. A party can reveal the information they contain by turning over the documents to the other party.

Recent changes have been made in the discovery rules in the federal courts. Under the old rules, parties had to disclose only those documents that the other side requested, but there was, as a practical matter, no limit on the number of requests that the other side could make. The new rules require—in the absence of a local rule to the contrary—parties to supply each other at the start of the case with all documents in their possession "relevant to disputed facts alleged with particularity in the pleadings." We want to ask how this change might affect the dynamics of litigation.

As is always the case when we confront verifiable information, we have to begin with the unraveling result. A simple model of private verifiable information suggests that discovery rules should not matter because the party with the most favorable private information will reveal it, those whose information is only slightly less favorable will therefore also reveal it, and eventually all will reveal their private information. The conditions under which unraveling takes place, however, do not always hold. Here, as elsewhere, we need to begin by identifying the barriers that prevent unraveling and then assess their implications for law reform. When we are looking at the old rule, under which a litigant had to ask for specific information, we see that the most important barrier might have been that the litigant did not know whether the other had any relevant information.

Assume that Victim suffered damages from a product that was bought from Injurer and that the outcome of the litigation will turn on whether the product poses serious health risks. Injurer may have a great deal of information about this product that would be relevant to the outcome of the litigation, or Injurer may have none at all. A memorandum written by one of Injurer's research scientists may discuss experiments in Injurer's laboratory that reveal that Injurer's product poses an enormous health risk. There is some chance, however, that no internal research was done on the health effects of the product. Let us take a very simple example of this phenomenon. Victim believes that there are three possibilities, each equally likely: research was undertaken and the results indicated that the product was safe; research was undertaken and the results indicated that the product was unsafe; or no research was undertaken. Assume that, given the state of the other evidence, Victim prevails only if Injurer did the research and discovered that the product was unsafe. The information will never come out unless Injurer discloses it. Under the old rule, Victim might have had incompetent or unlucky lawyers who failed to ask for the records of any research that Injurer might have done. Moreover, the case might have been dropped if initial discovery proved fruitless.

The unraveling result suggests that Injurer will always disclose the information, but this result would not hold here. Injurer would, of course, disclose if the research showed that the product was safe; Injurer would not disclose, however, if the research showed that the product was unsafe. This type of injurer could hide because the injurer who did not research would have nothing to disclose. When an injurer failed to disclose, we could infer only that the outcome of any research that was done was not favorable. We could not infer that any research was in fact done. In this example, we would know only that there is a 50 percent chance that an injurer who did not disclose was one who did research that showed the product was unsafe.

Injurer may have had other reasons for not voluntarily disclosing information at the outset. The uninformed party may not have realized that the informed party had verifiable information and therefore did not make a negative inference from silence. If the informed party was reasonably sure that the uninformed party would fail to draw the correct inference, unraveling would again not occur. If Injurer, for example, had private information that was favorable, Injurer may have been better off waiting to disclose it so that the other side would be less able to respond to it effectively.

Revelation of information is also costly. Injurer's scientists may have done experiments that proved the product was safe, but it may have been hard to convey these results to Victim or Victim's lawyers. If the uninformed party did not know how large the revelation costs were, complete unraveling would have been unlikely. Finally, even the informed party with favorable information may not have known that the information was favorable. Research, for example, may have shown that the product was dangerous only for persons with a specific, preexisting condition. Whether Injurer disclosed this information, however, depended in part on whether Injurer knew about Victim's condition.

We now consider how we might model discovery rules in situations in which one side may have private information that is verifiable— information that a party could credibly reveal if it chose to do so. Such a model could help us identify some of the inefficiencies under the old federal discovery regime and suggest whether the new regime in which disclosure is mandatory is desirable.

Let us assume that Injurer prevails unless Victim acquires private information that at the outset only Injurer knows. Injurer's private information may increase Victim's chances of prevailing on the merits only a small amount or a great deal. Injurer may reveal the information voluntarily, but even after Injurer discloses some damaging information, Victim can never know whether Injurer has other information that is even more damaging. Revelation of a lab report suggesting that the product poses only a small health risk does not exclude the possibility that a follow-up study showed an even greater risk. Thus, voluntary disclosure of some information does not lead Victim to call off efforts to gain additional evidence through discovery. Victim has a chance of winning only by gaining evidence, either through discovery or through voluntary revelation.

We can model the interaction between Victim and Injurer in the following way. Injurer at the outset has a single chance to decide whether to disclose private information. Injurer can also make a settlement offer, which Victim can accept or reject. If the offer is rejected, Victim makes a discovery request, choosing to spend a certain amount on discovery. Every dollar that Victim spends on discovery inflicts a cost on Injurer of twice as much. Discovery is imprecise. For any finite expenditure on discovery, it is not certain that Victim will ask for the damaging evidence. We can model this by assuming that the probability of finding the evidence increases as Victim's expenditures on discovery increase. By contrast, voluntary disclosure of information that increases Victim's likelihood of succeeding is costless.

These assumptions formally capture the idea that Injurer knows the location of the smoking gun, but Victim must search through thousands of documents and dozens of depositions to find it. If discovery is successful, it uncovers all the damaging evidence; if it is unsuccessful, it uncovers no damaging evidence. After discovery, Injurer can make another settlement offer. To simplify this stage of the game, we assume that Injurer pays the expected value of the trial outcome given the information that Victim has acquired after voluntary disclosure and discovery.

Let us examine Injurer's choices at the start of the game. Injurer may make a settlement offer that signals to Victim that Injurer is in fact liable. Sending this signal, however, is different from actually disclosing the information. If Injurer makes an offer that Victim rejects, Victim must still spend resources on discovery to obtain the information. Settlement offers are not admissible as evidence. Hence, Victim cannot argue in court, "Injurer's settlement offer was so large that it shows that Injurer is liable." In contrast, once Injurer discloses the information, Victim both learns something about the probability that Injurer will be found liable and no longer has to spend any resources on discovery to acquire that information for use at trial.

The possibility that Victim can save resources on discovery may make settling with Injurer attractive. These resources can be saved in either of two ways: Injurer can make a settlement offer that conveys information to Victim, or Injurer can disclose the information outright. Victim cannot use the information signaled through an offer in the litigation, but Victim can take advantage of any information voluntarily disclosed.

In the equilibrium of this game, Injurer discloses nothing. Injurers will act in one of two ways. One type of injurer, those with the more damaging evidence, will make a settlement offer before discovery, and that offer will be accepted. Injurers who are the most likely to be found liable do not disclose any private information because it eliminates the possibility that Victim will learn nothing in discovery. The other type, of injurer, those without damaging evidence, will make no offers, reveal no information, and expose themselves to discovery. Injurers who have only a low probability of being held liable cannot disclose their type to Victim, because injurers with a high probability of being found liable can mimic them by disclosing the same information, but not the information that is most incriminating. Thus, voluntary disclosure does no good for the injurers with only a low probability of being held liable either.

Unlike the other type of injurer, those with a low probability of being held liable are not willing to make a large settlement offer. Therefore, they remain silent and go through discovery. In this model, injurers most likely to be held liable pool with injurers who are least likely to be held liable, as neither discloses information voluntarily. Those who are least likely to be held liable go through discovery and, at the end of the day, incur costs greater than their ex ante expected liability.

Mandatory disclosure may improve matters significantly. It is somewhat difficult to model mandatory disclosure explicitly. One needs to incorporate a monitoring technology and sanctions for failure to disclose if monitoring indicates that a party concealed information. If the monitoring technology is good enough, Victim may have to commit to spending only modest resources monitoring compliance with the disclosure requirements to ensure that all injurers have an incentive to disclose. Of course, there are a number of practical problems with a mandatory disclosure regime. The rules must set out exactly what kind of information must be disclosed. How these rules apply in particular cases will itself become a subject of litigation. Victims may initiate a lawsuit and impose costs on injurers in the hope that they may pick an injurer whose private information is devastating. This possibility introduces a potentially large inefficiency. The process for monitoring compliance with the rules must be established in a way that prevents abuses similar to those in the current discovery process. Nonetheless, the model points out a number of reasons the old discovery process might not have worked well and indicates why a mandatory disclosure regime might be an improvement.

Summary

In this chapter, we focused on the way in which problems in information affected legal rules governing bargaining. This is a large issue, but we narrowed our focus to these problems in the context of litigation and saw a general theme emerge. Those individuals least likely to be held liable, those against whom rights are the weakest, may be the ones who must incur the highest costs in order to distinguish themselves from others. The basic structure of these models, however, extends beyond problems in civil procedure. They are useful paradigms for examining information problems that arise in corporate reorganizations, labor, and commercial law.

Bibliographic Notes

Basic models of the litigation process.   The optimism model of settlement and litigation was developed by Gould (1973), Landes (1971), and Posner (1973). One-sided private information models of litigation and settlement include Bebchuk (1984), Katz (1990a), Nalebuff (1987), Reinganum and Wilde (1986), and P'ng (1983). Snyder and Hughes (1990) and Hughes and Snyder (1991) contain empirical analyses of settlement and litigation under the English and American rules. The results illustrate the contrast in the implications of the optimism and private information models. The solution concept that is used to solve the sequential trial model with private information is the *Farrell, Grossman, and Perry refinement* of the perfect Bayesian equilibrium. See Farrell (1983) and Grossman and Perry (1986). See Kennan and Wilson (1989) for an information-based empirical analysis of strike data. Daughety and Reinganum (1991) analyzes a model with endogenous timing. The dynamics of settlement are analyzed by Spier (1992).

The differences between litigation models and the more standard model (bargaining with private reservation values) are discussed in the mechanism design analysis of Spier (1989) and Spulber (1990). They show that there are no efficient mechanisms for the one-sided private information model. This is in contrast to bargaining with private reservation values, where it is possible to achieve efficiency with one-sided private information but not with two-sided, as shown by Myerson and Satterthwaite (1983). See Vincent (1989) for a nonlitigation model of bargaining with common values.

We have suggested that when one side has information that another does not, it may be rational for one party to wait in order to distinguish itself from others. We must be careful, however, not to press this point too hard. Coase (1972) suggests that a monopolist who sells a durable good will set a price that approaches the competitive price as the period between sales becomes arbitrarily short. This idea is known as the *Coase conjecture*. One can model the offers that an uninformed party makes to the informed party in the same way to suggest that, as the period between offers becomes shorter, agreement between the parties takes place earlier and the gains to the informed party approach $0. For a discussion and formal proofs, see Gul, Sonnenschein, and Wilson (1986) and Gul and Sonnenschein (1988). For an examination of the Coase conjecture's importance for game theory, see Fudenberg and Tirole (1991a).

Experiments that focus on litigation and information are Loewenstein, Issacharoff, Camerer, and Babcock (1993) and Babcock, Loewenstein, Issacharoff, and Camerer (1992).

Information and selection bias. The classic paper on the selection of lawsuits that are tried is Priest and Klein (1984), which shows that, if the parties have equal litigation costs, equal stakes, and differ only in their beliefs about liability, the outcome of litigated cases will tend toward an even split between plaintiff and defendant victories. Further papers include Priest (1985), (1987), Eisenberg (1990), Wittman (1985), (1988), and Cooter (1987).

Sequential trials. The optimism version of the bifurcated trial model is a special case of Landes (1993). The unitary version of the private information model of the bifurcated trials is very similar to Sobel (1989). Another litigation model with two-sided private information is Schweizer (1989).

Discovery rules and verifiable information. The only litigation model with verifiable information that we know of is Shavell (1989). Economic analyses of discovery rules are contained in Setear (1989) and Easterbrook (1989).

Information and the Limits of Law

The tools of game theory, like the tools of economics more generally, are not powerful enough to tell us the precise effects of particular laws. Nevertheless, we are not completely at a loss, for the formal tools of game theory offer us a starting place for understanding the forces that are at work in situations in which parties behave strategically.

We have shown that legal rules often affect parties in ways that are largely invisible. In many instances, a legal rule has dramatic effects even though it attaches consequences to actions that individuals never take, regardless of whether the legal rule is in place or not. By attaching consequences to this combination of actions, a legal rule leads individuals to shift from one combination of actions to another. The models that we have set out show how the forces that determine outcomes when parties interact strategically often work beneath the surface, by attaching consequences to actions that are not taken before or after the legal rule is put in place. We cannot gain the right set of intuitions for how laws work unless we first know where and for what we should be looking: We must look not only at the consequences that the legal rule attaches to actions parties take, but also at the consequences that the legal rule attaches to actions that parties might not take even in the absence of the legal rule.

Laws can be enforced only to the extent that a court has access to the information needed to implement them. When analyzing the behavior of parties in situations in which one or both has information that the other does not, we have to be especially careful in identifying the kind of information that we are confronting. Indeed, sorting out different types of information and the way legal rules affect them is one of the most distinctive characteristics of work that applies the tools of game theory to legal analysis. The analysis we have offered in this book provides a framework for understanding any legal rule that regulates the transmittal of information. These legal rules range from disclosure rules governing the sale of a home to various antidiscrimination laws.

The context in which legal rules operate matters. The purpose of using economic tools to analyze legal problems is to build simple models that capture the forces at work. This is a particularly difficult task when we are confronting situations in which parties behave strategically. Often one cannot easily look at an interaction in isolation. In many situations, for example, parties have repeated interactions of the same kind. Behavior that would not be sustainable without repeated dealings becomes plausible, thereby enlarging the set of problems that laws may need to address. Repetition is not the only way in which context matters, however. A particular interaction may occur within a much larger web of interactions. When a small game is embedded in a much larger game, laws that might seem sensible in the isolated small game appear insufficient when considered in the larger context.

It is well recognized that legal rules often merely set the ground rules for negotiations. Much interaction between individuals in the marketplace and elsewhere takes the form of bargaining in the shadow of the law. To understand how legal rules work, one must also understand the dynamics of bargaining. The formal tools of game theory thus also help us to understand the give-and-take of bargaining better.

When we are concerned with the special role that legal rules play for individuals who are largely free to structure their own affairs, we must be aware of how this ability limits the way legal rules can, in fact, alter individual behavior. Indeed, in contexts in which transaction costs are low and the forces of competition strong, the problem of strategic behavior may be minimal and the role of law merely interstitial. In those areas in which the problem of strategic behavior looms large, however, one must be exceedingly careful in defining the contours of the problem. In few contexts is it easy to isolate a particular strategic interaction from all others. It is our hope that the way we confronted these problems will again offer help to both lawyers and economists.

In the first instance, game theory helps us to compare different laws. It allows us to see, for example, that regimes of negligence, comparative negligence, and strict liability coupled with contributory negligence work in the same way and are, in fact, different variations on the same general rule. As long as the parties bear their own costs of care and are never liable for more than their costs of care when they exercise care and others do not, they have the incentive to act efficiently. All the regimes we see in Anglo-American tort law share these features, and they alone drive the familiar observation that, under strong assumptions, these regimes all induce efficient outcomes.

In the first chapter of this book, we showed that, in the absence of information problems, a legal regime of civil damages is enormously powerful. A set of rules under which one party has the right to recover money from another when a legal norm is violated can, in theory, ensure that each individual acts in a way that is in the joint interest of everyone. The socially optimal outcome can be reached as long as we accept the idea that players will choose a particular strategy when that strategy strictly dominates all others.

There is substantially more to game theory than this, however. Simple models of strategic interaction depend upon unrealistic assumptions about information. The central challenge facing those who want to know how legal rules affect behavior is understanding how to craft legal rules in a world in which information is incomplete. In many cases, the necessary information is not available to the court or is lacking altogether. In other cases, only one party possesses information, and that party may have neither the means nor the inclination to disclose it. The legal analysis of problems involving strategic behavior requires us to solve a puzzle in which many of the pieces are either missing or in the possession of people who will not or cannot give them up.

The effectiveness of legal rules turns on whether the rules take into account what the parties themselves know and whether the demands they place on the courts are reasonable. The limited ability of parties as well as the courts to determine what behavior constitutes due care constrains the ability of a legal rule to induce everyone to exercise such care. A law cannot prevent lying when a court cannot determine whether someone is telling the truth. Taking account of who possesses what information is a necessary step in understanding a legal problem in which parties behave strategically.

For too long, distinctions between verifiable and nonverifiable information and questions such as whether information is observable only to the parties have been neglected. Disclosure laws, for example, are often criticized without any attempt to confront whether the information is verifiable, and, if it is, why unraveling does not take place. The tools we have introduced in this book allow us to confront these problems rigorously and understand whether legal rules make the correct assumptions about what the parties know and what the courts can learn.

Laws matter because they affect the way people behave. To understand these effects, we must have some means of predicting behavior. Game theory posits that, regardless of how people go about making decisions, the actions they take are consistent with a few basic principles. These include the idea of strict dominance. An individual is likely to choose an action if that action leaves that player better off than any other action would, regardless of what other players do. Another principle is the one on which the Nash equilibrium concept rests: If an interaction has a predictable outcome, the outcome is likely to be one in which no one can do better by choosing another action, given the actions others take.

The power of game theory as applied to law or any other discipline must derive in the end from the accuracy of these predictions about the choices people make. Precisely because these predictions can be wrong, they are also capable of being right and are therefore testable. They can help us to answer questions that are important to everyone who cares about the law. As we begin this enterprise, we need to remember that these principles must be scrutinized as carefully as the legal questions themselves. Such scrutiny, however, can take place only after we try to apply them. Understanding the law and understanding game theory are challenges that go hand in hand.

Notes / References
Glossary / Index

Notes

1. Simultaneous Decisionmaking and the Normal Form Game

1. In this example and elsewhere in the book, we assume *risk neutrality*. This assumption in no way changes the character of the results, provided that we can aggregate the utility people derive from different states of the world when there is some probability that each of these states will arise. We can in fact make such aggregations if the axioms of *von Neumann-Morgenstern utility* hold. In the bibliographic notes to this chapter, we point to sources that spell out in detail von Neumann-Morgenstern expected utility theory and other assumptions fundamental to economic analysis. In the main, these assumptions seem both plausible and consistent with observed economic behavior. At this point, we want to note only that these assumptions are fundamental to the analysis and, though reasonable and generally accepted, are not self-evident. There is, for example, a thriving literature on "anomalies"—experiments that suggest that these assumptions do not always hold.

2. Game theory has explored extensively strategies that are dominant but not strictly dominant. Our focus, however, is on strictly dominant strategies, and, unless otherwise noted, we shall use "dominant" as a shorthand for "strictly dominant."

3. Among other things, this observation about the motorist's incentives is independent of the payoffs we use in this model. It holds whenever one assumes that people do not play dominated strategies, that motorists do not bear the full costs of an accident, and that taking care is costly.

4. California Book of Approved Jury Instructions No. 14.91 (7th ed. 1986).

5. Michigan Standard Jury Instructions 2d §42.01 (1991).

6. See Schwartz (1978).

7. New York Civil Practice Law and Rules §1411.

8. This concept is also sometimes called a "Cournot-Nash equilibrium."

Cournot was a nineteenth-century economist whose analysis of oligopolistic competition used the same concept in a less general form.

9. Kreps (1990a), p. 135, offers the following game as an example:

> There are two players, A and B. The game has either one or two stages. In the first stage, A and B simultaneously and independently choose between the letters X and Y. If both choose Y, both receive $1 and the game is over. If one chooses X and the other chooses Y, both receive $0 and the game is over. If both choose X, the game proceeds to a second stage. In this second stage, the two simultaneously and independently choose positive integers. These integers are then compared and payoffs are made; if the integers chosen are different, the player choosing the higher integer wins $250, and the player choosing the lower integer wins $100. If the integers are the same, each player wins $25.

In this game, the Nash equilibrium concept suggests that both players choose Y in the first stage. If the other player is playing Y, the best response is to play Y as well. No other strategy combination can be Nash. A strategy combination in which both players play X and choose an integer can never be Nash, given that one or the other player will have the lower integer and could have done better by picking the higher integer. The player with the lower integer will not be playing a best response given the strategy choice of the other player.

10. Landes and Posner (1980) offers the first proof of this proposition in the context of an analysis of joint tortfeasors.

11. The proof for the two-player game is as follows. Assume that a lawmaker wants Player 1 to adopt strategy x^* and Player 2 to adopt strategy y^*. The lawmaker can provide first that Player 1 must pay D in damages to Player 2 if Player 1 adopts any strategy other than x^*. Second, the lawmaker can provide that, whenever Player 1 plays x^*, Player 2 must pay damages D, unless Player 2 plays y^*. Finally, the lawmaker provides that, if the players choose x^* and y^* respectively, neither owes any damages to the other. The civil damage award D is some amount that is greater than the difference between any two payoffs for any player for any given strategy before the liability rule is put in place. When this civil liability regime is adopted, the strategy combination (x^*, y^*) is the solution to the game. Player 1 plays x^* because it strictly dominates all other strategies. (D is large enough so that Player 1 is always worse off having to pay damages D than not having to pay them.) Given that Player 2 believes that Player 1 will not play a dominated strategy, Player 2 believes that Player 1 will play x^*. Player 2 is therefore better off playing y^*. This strategy is the only one that Player 2 can play and avoid paying damages D, given that Player 1 is going to play x^*. (The sum of D and any other y exceeds y^*.)

An infinite number of civil damages regimes must exist, given that the same proof works when damages are set at nD, where n is any number greater than 1. The proof for the n-player game takes the same basic form.

The problem in crafting a civil damages regime is in finding the legal regime among all those possible that makes the most sense, given such things as the cost of implementing a legal regime and solvency constraints.

Nothing in this proof requires (x^*, y^*) to be Pareto optimal. As we show in the next section, however, we must inject such a requirement in order to have a civil liability regime that makes the strategy that we want parties to adopt one that strictly dominates all others.

12. This problem is discussed in In re Sutter-Butte By-Pass Assessment No. 6, 191 Cal. 650, 655–656 (1923). There are now many statutes regulating the building of levees, and these displace the common law rule to some extent in many jurisdictions.

13. Id. at 656.

14. Closely related to the stag hunt is the "assurance game." In the assurance game, both parties prefer to cooperate with each other and engage in a common enterprise (such as hunting stag). If one player decides to hunt hare instead, however, that player is better off if the other player hunts stag. (Hunting hare may have a payoff of $8 when the other hunts stag, but only $6 when the other player hunts hare as well.) The levee example would become an assurance game if the first landowner's decision to maintain the levee brought the other a benefit, even if the other player did not maintain the levee. It is called an assurance game because even after one decides not to engage in the cooperative strategy (maintaining the levee or hunting stag), one still has an incentive to pretend that one is.

15. We can also calculate the expected payoffs in this way: $(0.5 \times -4) + (0.5 \times -10) = -7$.

16. To put the point algebraically, we must solve for p_1, where $(-4 \times p_1) + [-10 \times (1-p_1)] = -6$.

17. Schelling (1960).

18. See Cooper, DeJong, Forsythe, and Ross (1990).

19. For an example of such a rule, see Hayashi v. Alameda County Flood Control and Water Conservation District, 334 P.2d 1048 (1959).

20. Such a strategy on the part of management is known as *Boulwareism*, after the General Electric executive who first promoted it. The general rule is that "an employer may not so combine 'take-it-or-leave-it' bargaining methods with a widely publicized stance of unbending firmness that he is himself unable to alter a position once taken." See NLRB v. General Electric Co., 418 F.2d 736, 762 (2d Cir. 1969).

2. Dynamic Interaction and the Extensive Form Game

1. 96 F.T.C. 653 (1980).

2. 771 F.2d 1093 (7th Cir. 1985).

3. This schedule raises an issue that we confront directly in the next chapter. Because the schedule is tied to the actions that the parties take, we are

necessarily assuming that a court is able to determine after the fact what kind of test Dairy used. This assumption may seem odd. At the time Processor decides whether to test, it cannot tell how Dairy tested. If Processor cannot acquire this information, it might seem that a court could not either. A court, however, may be able to gather this information, given that in litigation documents can be discovered and witnesses can be cross-examined under oath.

4. We could have also solved this subgame using dominance solvability. (Dairy's strategy of high now dominates low. Processor, knowing that Dairy will therefore play high, will play medium.) When a single strategy combination remains after the iterated elimination of dominated strategies, this strategy combination is also the unique Nash equilibrium to this game.

5. The relevant provisions of the Uniform Commercial Code for the implied warranty of merchantability and the duty to mitigate are 2–314 and 2–715 respectively.

6. See Shavell (1980).

7. See Epstein (1989).

3. Information Revelation, Disclosure Laws, and Renegotiation

1. 15 U.S.C. §§78a et seq. (Williams Act).

2. 29 U.S.C. §§2101 et seq. (Worker Adjustment and Retraining Notification Act).

3. If Processor tests low, it receives $(0.5 \times 3) + (0.5 \times 0) = 1.5$. If Processor tests high, it always receives a payoff of 1.

4. Processor is better off testing low given this belief, because getting a payoff of $3 40 percent of the time is better than getting a payoff of $1 all the time.

5. 380 U.S. 609 (1965).

6. 450 U.S. 288 (1981).

7. 450 U.S. at 303 (footnote omitted).

8. See 42 U.S.C. §12112(c)(2)(A).

9. See 34 CFR §104.42(b).

10. See 24 CFR §100.202(c).

11. See 10 CFR §1040.31, 1040.56, 34 CFR §106.21, 106.60, 45 CFR §86.21, 86.60.

12. See, e.g., In. St. 4–15–2–15; NJ St. 18A:6–5; NY Civ Serv App 6.2; NY Civ Rts 40-a.

13. Il St Ch 48 ¶2860 (effective July 1, 1992).

14. See, e.g., Federal Rules of Evidence, Rule 412.

15. See Model Rule of Professional Conduct, Rule 5.6(b).

16. Id.; see also Model Code of Professional Responsibility, DR 2–108(b).

17. 351 U.S. 149 (1956).

18. See Kennan and Wilson (1990).

19. We explore this point in an explicit model below.

20. See California Civil Code 1102–1102.15.
21. As we discuss in the bibliographic notes, this model builds on Shavell (1991).
22. 15 U.S. (2 Wheaton) 178 (1817). Cooter and Ulen (1988) emphasizes the need to take account of whether information has social value in assessing legal rules.
23. $(\frac{1}{2} \times 200) + (\frac{1}{2} \times 180) = 190$.
24. In other words, the silent seller can expect to get $(\frac{2}{3} \times 180) + (\frac{1}{3} \times 200) = 186.67$.
25. The pool of silent sellers contains $\frac{3}{4}$ of all sellers: $\frac{1}{4}$ with high costs and well-built houses worth $200, $\frac{1}{4}$ with high costs and badly built houses, and $\frac{1}{4}$ with low costs with badly built houses. Because the chances are 2 in 3 that the house is worth $180 rather than $200, a buyer would be willing to pay $186.67 to everyone in this pool.
26. The low-cost seller expects to receive $(0.5 \times 205) + (0.5 \times 185) - 4 = 191$.
27. The high-cost seller expects to receive $(0.5 \times 205) + (0.5 \times 185) - 8 = 187$.
28. One-third of all silent sellers have well-built houses. The probability of finding one on the first search is $\frac{1}{3}$. The probability of not finding one on the first search but finding one on the second is $\frac{2}{3} \times \frac{1}{3}$. The probability of not finding one on the first or second but finding one on the third is $(\frac{2}{3})^2 \times \frac{1}{3}$. The expected number of searches is $[(\frac{1}{3}) \times 1] + (\frac{2}{3} \times \frac{1}{3} \times 2) + [(\frac{2}{3})^2 \times \frac{1}{3} \times 3] + [(\frac{2}{3})^3 \times \frac{1}{3} \times 4] + \ldots = 3$.
29. Half the sellers gather information. Hence, sellers alone produce .5 searches per house. Half the buyers who value well-built houses will buy them from sellers who reveal that they have well-built houses. The remaining buyers who want to buy well-built houses (or $\frac{1}{4}$ of all buyers) will have to search. As we show in note 28, each of these buyers can expect to make 3 searches. Hence, the total amount of information gathering is $\frac{1}{2} + (\frac{1}{4} \times 3) = 1.25$ per house.
30. For a discussion of common knowledge, see Geanakoplos (1992).
31. 117 F. 99 (1902).
32. See Joskow (1985).
33. For a discussion of incomplete contracts and problems associated with contracts that are insufficiently state contingent, see Ayres and Gertner (1992).
34. As we will see, the asymmetry in the investment functions (Buyer's investment technology is more efficient than Seller's) is essential for an incomplete contract to do better than no contract in this setting.
35. Because trade is always socially efficient (Buyer's value exceeds marginal cost), the social optimum is found by choosing k_b and k_s to maximize $E(v) + k_s + k_b - 2k_s^2 - k_b^2$.
36. An even division is consistent not only with intuition but also with formal bargaining models, such as the Rubinstein bargaining model that we discuss in detail in Chapter 7.

37. We should note, however, that the optimal contract price will turn on it.
38. Solving for the optimal price is straightforward: First, one derives expressions for expected surplus to each side as a function of investments and contract price. Then one solves for the Nash equilibrium of the investment game, for any contract price. One takes the solution to this game and calculates social surplus as a function of the contract. Finally, one chooses the contract price to maximize equilibrium social value.
39. 499 F. Supp. 53 (W.D. Pa. 1980).
40. For an opinion that discusses the way in which parties use terms in their contract as a baseline for future renegotiations, see PSI Energy, Inc. v. Exxon Coal USA, Inc., 991 F.2d 1265 (1993).

4. Signaling, Screening, and Nonverifiable Information

1. See Elster (1989) at p. 118.
2. The interaction set out in Figure 4.1 is generally known as a *beer-quiche game*. Like other paradigms in game theory, it is based on an offbeat story. In this case, the story takes the following form. One player is a bully and the other player is a patron of a restaurant. The patron is one of two types (a tough guy or a wimp). The patron must decide what to order for breakfast (beer or quiche), and the bully must decide, after seeing what the patron orders, whether to pick a fight. Neither tough guys nor wimps like to fight. The former prefer beer for breakfast, the latter quiche. Bullies like to fight wimps, but they do not like to fight tough guys. In our game, the bully corresponds to the employer, the restaurant patron to the worker, and the choice of beer or quiche to the choice of the swivel or the orthopedic chair.
3. It goes without saying that these background beliefs affect the play of the game. We would, for example, have a completely different equilibrium if 10 percent of the workers had good backs and the rest had bad backs. In this situation, the only perfect Bayesian equilibrium to survive refinements is a partially separating equilibrium. Workers with bad backs ask for swivel chairs $\frac{1}{9}$th of the time and the employer trains $\frac{1}{2}$ the workers who ask for such chairs.
4. The arithmetic here is: $(0.9 \times 1) + [0.1 \times (-1)] = 0.8$.
5. Refinements of this sort to the perfect Bayesian equilibrium concept are also called *Cho-Kreps refinements* after the economists who first set out the basic beer-quiche paradigm.
6. We add the benefits to both the workers and the employer when the workers have good and bad backs and take account of their likelihood: $0.1 \times (2 - 1) + 0.9 \times (3 + 1) = 3.7$.
7. We make the same calculation as before: $0.1 \times (3 - 1) + 0.9 \times (3 + 1) = 3.8$.
8. Again, we calculate: $0.1 \times (3 - 1) + 0.9 \times (2.9 + 1) = 3.71$.

9. We are assuming here that legal rules requiring or banning certain actions are enforced through fines or criminal sanctions. We discuss the effect of damage rules later.

10. These issues are discussed in both Posner (1992) and Kaplow (1992).

11. Family and Medical Leave Act of 1993, P.L. 103–03 (February 5, 1993).

12. Models in which both players have private, nonverifiable information are much more complicated. We introduce one in Chapter 8.

13. See, e.g., the Worker Adjustment and Retraining Notification Act, 29 U.S.C. §§2101 et seq.

14. A conspicuous exception is the refusal to enforce penalty clauses.

15. 9 Ex. 341, 156 Eng. Rep. 145 (1854).

16. A miller in the position of Baxendale today would go not to any shipper, but rather to one who specialized in overnight delivery. All the form contracts of these shippers provide that the shipper is not liable for any consequential damages in the event of a delay, whether reasonably foreseeable or not. Given that this market is one in which there is competition and in which those to whom it caters are sophisticated, one might well conclude that excluding consequential damages altogether is in fact an efficient outcome.

17. A second interpretation of foreseeable damages is that the carrier is liable for the minimum of actual damages and the average level of damages. This would imply liability of $100 for low-damage millers and $110 for high-damage ones. We adopt this interpretation in the text, but our argument does not depend on which of the two we use.

18. 382 P.2d 109 (Okla. 1962), cert. denied, 375 U.S. 906 (1963).

19. The information problem is in fact a subtle one, because the likelihood that Garland performs turns not only on its type, but also on the steps it takes during the course of performing the contract.

20. Implementing penalty defaults, however, is not necessarily easy. To work, a penalty default must lead parties to draft clauses that give them what they want, and drafting appropriate contract language is never easy. In the case itself, it appears that the Peevyhouses gave up the liquidated damages clause typically included in such contracts and explicitly negotiated for Garland's promise to restore the land. For a comprehensive analysis of *Peevyhouse*, see Maute (1993).

21. See Maute (1993).

22. Giving the Peevyhouses a right of specific performance, however, does not itself necessarily lead to economic waste. Garland and the Peevyhouses can negotiate with each other, in which case the Peevyhouses might be willing to accept cash from Garland and in turn release it from its obligation to restore the land. Whether these negotiations would in fact lead to such a settlement, however, is not clear. We pursue this question in Chapter 7.

23. In this example, we are making things simpler than they are. The optimal

amount of insurance is an amount that gives someone the same marginal utility in all states of the world. It is possible that there are some states of the world in which you would value money less (because, for example, you were in a permanent coma). If such states existed, you would want insurance that gave you less wealth in that state. What matters is that the value you put on the next dollar that you acquire or spend has to be the same in all states.

5. Reputation and Repeated Games

1. American Tobacco Co. v. United States, 328 U.S. 781 (1946).
2. This implies that the interest rate is $(1 - \delta) / \delta$. If the interest rate is r, the present value of a dollar in the subsequent period is $1 / (1 + r)$. Equating $\delta = 1 / (1 + r)$ gives $r = (1 - \delta) / \delta$. A perpetuity of K dollars each period would imply a present value of $K / (1 - \delta)$ dollars in period 1, because $K + \delta K + \delta^2 K + \delta^3 K + \ldots = K / (1 - \delta)$.
3. 964 F.2d 335, 341 (4th Cir.), affirmed, 113 S. Ct. 2578 (1993).
4. 257 U.S. 377 (1921).
5. 268 U.S. 563 (1925).
6. Most-favored nation clauses are going to be in the interest of firms in an industry only if the others adopt them as well. A firm that does not have such a policy could adopt price cuts, and it might be too costly for the firms with such policies to respond.
7. See, e.g., Easterbrook (1981).
8. United States v. American Tobacco Co., 221 U.S. 106 (1911).
9. The word "rational" is commonly used in these games to describe the actions of someone whose payoffs are entirely monetary. We conform to that usage here. This use of the word, however, is radically different from the one we have been using throughout the book. Elsewhere, we use "rational" to capture the idea that individuals try to advance their own self-interest as they perceive it, regardless of whether they care only about money or about other things as well. In this sense, both kinds of incumbent in this game are rational.

6. Collective Action, Embedded Games, and the Limits of Simple Models

1. See Sunstein (1990).
2. See Roe (1987).
3. Security interests matter in this model because they create priorities among creditors. Firms that have only one creditor, as many do, do not face any of the problems we discuss in the text.
4. We can calculate the payoffs of a creditor who does not monitor when the other one does by the following calculation: $(.8 \times 52) + [.2 \times (84 - 52)]$

= 48. When both creditors monitor, each receives an expected return of $50, less their costs of monitoring, or $45 and $42 respectively.

5. Together, the two creditors lend $100 and incur costs of $13 (or $113 in all), but they receive only $84 when the project fails. To make up for the shortfall when the project fails, they must be paid an amount large enough when the project succeeds so that their expected return is $113. This amount is $142. We can see this by noting that $(0.5 \times 142) + (0.5 \times 84) = 113$. Creditor 1 must expect to receive $55 back from the debtor. (It must recover its capital and the $5 it expects to spend on monitoring.) Creditor 2 must expect to receive $58. When the project fails, Creditor 1 and Creditor 2 receive a share of $84 proportionate to the amount of their original loan.

From this information, we can calculate how much the debtor promises to pay each creditor. As noted, the total amount that the debtor is obliged to pay is $142, and this amount is divided between Creditor 1 and Creditor 2 in some way. We can call the portion that Creditor 1 receives s_1 and the portion that Creditor 2 receives s_2. They share in the $84 that the debtor has when the project fails in the same proportion. We can therefore calculate the amount the debtor promises to pay each by solving the following:

$$(0.5 \times 142 \times s_1) + (0.5 \times 84 \times s_1) = 55$$
$$(0.5 \times 142 \times s_2) + (0.5 \times 84 \times s_2) = 58$$

When we do this, we find that the debtor promises to pay Creditor 1 $69 and Creditor 2 $73.

6. This is an example of a *direct mechanism*, or a game in which the only action of the players is to announce their type (the value they place on the landscaping), as opposed to some other allocation mechanism in which the players take some other action (such as promising to contribute a certain amount of money for the landscaping) that might reveal their type. We focus on direct mechanisms because if any allocation mechanism leads to a particular equilibrium, there also exists a direct mechanism in which the players truthfully reveal their type that implements the same equilibrium. If no direct mechanism in which parties truthfully reveal their type exists that reaches a particular equilibrium, no other scheme we can devise can reach it either. This theorem—called the *revelation principle*—is useful in modeling public goods, auctions, regulation, and any other environment in which the problem is one of designing an allocation mechanism in the presence of private, nonverifiable information. See Gibbons (1992), Myerson (1991), or Fudenberg and Tirole (1991a) for formal treatments.

7. It is only a weakly dominant strategy because for some distributions of value a player would be no worse off stating a value different from the player's true value.

8. In this game and the two that follow, we have rounded numbers for sim-

plicity of presentation. This makes it appear, for example, that Player 2 receives the same payoff, $9, both from being on the old standard with Player 1 and the existing players and from being on the old standard with just the existing players. This is not the case, as the payoff from the former, by construction, exceeds the latter.

9. See, e.g., Computer Associates International, Inc., v. Altai, Inc., 998 F.2d 693 (2d. Cir. 1992) and the cases cited therein.

10. 799 F. Supp. 203 (D. Mass. 1992).

11. For a more detailed analysis of these issues, see Menell (1987, 1989).

12. The example follows the discussion in Banerjee (1992).

13. See Baer and Saxon (1949) at p. 157.

7. Noncooperative Bargaining

1. For proof that this is the unique subgame perfect equilibrium to this game (as well as the only one to survive repeated elimination of dominated strategies), see Kreps (1990b).

2. Determining the value of the share as the bargaining periods become arbitrarily short is a little harder than it appears because of the exponential form of the function δ. The natural logs appear in the equation on this account as well.

3. 382 P.2d 109 (Okla. 1962), cert. denied, 375 U.S. 906 (1963).

4. To be more precise, the Peevyhouses will make an offer that approaches the value of the exit option as the period between offers approaches 0.

5. Garland has a 50 percent chance of receiving nothing and a 50 percent chance of recovering ½ of the $1 million bond. Hence, its expected payoff is $250,000.

6. 11 U.S.C. §362. The Bankruptcy Code also gives Manager the exclusive right to propose a plan of reorganization for 120 days, but the exclusivity period does not affect the dynamics of the bargaining so long as the bankruptcy court is unwilling to confirm any plan that cuts off the rights of Manager.

7. 308 U.S. 106 (1939). In re Bonner Mall Partnership, 2 F.3d 899 (9th Cir. 1993), cert. granted, 114 S. Ct. 681 (1994). In other words, Manager may have what is in effect a call upon the stock of a reorganized Firm. Manager can exercise this right only by contributing cash to Firm that is equal to the value of the equity of the recapitalized company. This limitation may prevent Manager from enjoying the entire going-concern surplus. Given the new value that Manager must add, Manager can enjoy the surplus in the form of a wage that is higher than the alternative elsewhere, not in the value of the equity itself. There may be constraints (such as the wage ordinarily paid to a manager of such a firm) that limit how much money may be extracted from Firm in this form.

8. Creditor has this right because the two conditions of 11 U.S.C. §362(d) are

satisfied. The debtor has no equity in the property (under our assumptions), and, because there is no going-concern value, the property is not "necessary for an effective reorganization."

9. 406 U.S. 272 (1972).
10. 304 U.S. 333 (1938).
11. Note that this model suggests that the relative shares of the parties may shift if, for example, the union must wait longer between periods because it may have to obtain approval from its members before accepting or rejecting offers.
12. See Fernandez and Glazer (1991).

8. Bargaining and Information

1. Federal Rules of Civil Procedure 8.
2. Federal Rules of Civil Procedure 34(b).
3. Federal Rules of Civil Procedure 56.
4. Federal Rules of Civil Procedure 42(b) gives the court the power to order separate trials. For an example of such a case, see Polaroid v. Eastman Kodak, 641 F. Supp. 828 (D. Mass. 1985) and 17 U.S.P.Q.2d 1711 (D. Mass. 1990).
5. These amounts assume risk neutrality. If the parties are risk-averse, the settlement range will change somewhat, but the basic point is unaffected.
6. The calculations for Victim are $(0.9 \times 100,000) - 10,000 = 80,000$ and for Injurer $(0.6 \times 100,000) + 10,000 = 70,000$.
7. The use of "optimism" to describe these models is misleading. These models do not posit that parties are likely to be optimistic about their own chances. They assume only that parties are not always going to have the same view and that litigation arises when one or both of the parties happens to have a view that is too optimistic. Parties are equally likely to be pessimistic about their chances, but in these cases there is always settlement.
8. In some situations costless, nonbinding, and nonverifiable statements— or, in a phrase, "cheap talk"—made by Injurer may still signal information. See Farrell and Gibbons (1989) for such a bargaining model.
9. We take into account Victim's prior belief of likelihood of success and Victim's costs of litigating: $(0.5 \times 100,000) - 10,000 = 40,000$.
10. The calculations can be made as before: $(0.8 \times 100,000) + 10,000 = 90,000$, and $(0.2 \times 100,000) + 10,000 = 30,000$.
11. The calculation is $(0.5 \times 90,000) + 0.5 \times [(0.2 \times 100,000) - 10,000] = 50,000$.
12. Our model extends Landes (1993) by taking his basic model and assuming that each litigant possesses nonverifiable information.
13. Victim's expected recovery net of litigation costs is $(0.9 \times 110,000) - 10,000 = 89,000$. This amount is less than Injurer's expected costs of $(0.8 \times 100,000) + 10,000 = 90,000$.

14. The expected return to Victim if there is no settlement at any stage is $0.9 \times (110,000 - 5,000) - 5,000 = 89,500$, and the expected payment by Injurer is $0.8 \times (100,000 + 5,000) + 5,000 = 89,000$. The numbers in the text assume, however, that the parties settle after the liability phase in the event that Victim prevails on liability. After Victim prevails on liability, Victim's expected recovery (treating the costs of the trial phase as sunk) is $105,000 ($110,000 minus the costs of litigating damages), and the injurer expects to lose $105,000 ($100,000 plus the cost of litigating damages). Hence, the parties could settle for this amount. Whether they settle after litigating liability, however, does not affect the basic fact that litigation is more likely when there is bifurcation.

15. Bebchuk (1993) offers a model in which negative expected value suits are brought because the plaintiff incurs pretrial costs over time. Because the plaintiff is better off going through with the trial once pretrial costs are incurred, the defendant settles in the last period. Once the defendant knows that this will be the case, the defendant will settle in the next-to-last period as well. There is unraveling of the sort that we saw in Chapter 5 and the parties settle in the first period.

16. In a unitary trial regime, Victim's net recovery is $(0.9 \times 11,000) - 10,000 = -100 < 0$. In a sequential trial regime, Victim's expected recovery is $(0.9 \times (11,000 - 5,000)) - 5,000 = 400 > 0$.

17. To simplify the analysis, we do not allow an offer to stipulate damages and litigate liability under either regime.

18. We know the minimum offers that high- and low-damage victims accept respectively because $(0.75 \times 100,000) - 10,000 = 65,000$ and $(0.75 \times 60,000) - 10,000 = 35,000$.

19. Under the lower offer, which only low-damage victims accept, the expected payments are $(0.5 \times 35,000) + 0.5 \times [(0.75 \times 100,000) + 10,000] = 60,000$.

20. The careful injurer compares $(0.25 \times 80,000) + 10,000 = 30,000$ with $(0.5 \times 35,000) + 0.5 \times [(0.25 \times 100,000) + 10,000] = 35,000$.

21. We know the probability of winning on liability when a pooling offer is 0.5 because $(0.5 \times 0.75) + (0.5 \times 0.25) = 0.5$. We then know that the expected net recovery is $(0.5 \times 100,000) - 10,000 = 40,000$.

22. The careless injurer expects to lose $(0.5 \times 20,000) + 0.5 \times [(0.75 \times 100,000) + 10,000] = 52,500$ by making an offer of $20,000. In the event of litigation against both types of victims, the careless injurer expects to lose $(0.5 \times 0.75 \times 100,000) + (0.5 \times 0.75 \times 60,000) + 10,000 = 70,000$.

23. The careful injurer who litigates against everyone loses $(0.5 \times 0.25 \times 100,000) + (0.5 \times 0.25 \times 60,000) + 10,000 = 30,000$. By offering to settle for $20,000, the careful injurer expects to lose $(0.5 \times 20,000) + 0.5 \times [(0.25 \times 100,000) + 10,000] = 27,500$.

24. Injurer expects to lose $(0.5 \times 55,000) + 0.5 \times (100,000 + 5,000) = 80,000$.

25. If only high-damage victims litigate, careless injurers would expect to lose

in litigation $(0.75 \times 95{,}000) - 5{,}000 = 66{,}250$. (Victims accept a \$95,000 settlement offer after they win on liability.)

26. The expected costs to the careless injurer are the \$5,000 costs from litigating liability and \$80,000 in expected liability when confronting a pool consisting of both high- and low-damage victims, adjusted by the probability that Injurer will be found liable: $5{,}000 + (0.75 \times 80{,}000) = 65{,}000$.

27. High-damage victims must incur \$5,000 in attorney's fees for the liability phase. They have only a 50 percent chance of winning on liability. If they win, they must spend an additional \$5,000 litigating to receive their damages of \$100,000. (As noted, there is no equilibrium in which only high-damage victims litigate liability, and in which high-damage victims litigate in the damages stage when they are pooled with low-damage victims.) High-damage victims therefore expect to receive $-5{,}000 + (0.5 \times 100{,}000) - (0.5 \times 5{,}000) = 42{,}500$.

28. Once we recall that low-damage victims will settle after the liability phase for \$55,000, we find that $5{,}000 + (0.25 \times 0.5 \times 100{,}000) + (0.25 \times 0.5 \times 5{,}000) + (0.25 \times 0.5 \times 55{,}000) = 25{,}000$.

29. We have made all the calculations except for the expected return to the low-damage victim. Low-damage victims will always spend \$5,000 on litigation. They will prevail in the liability phase ½ the time and, if they prevail, will receive and accept a settlement offer of \$55,000. They can therefore expect $(0.5 \times 55{,}000) - 5{,}000 = 22{,}500$.

30. This argument was first advanced in Priest and Klein (1984).

References

Abreu, D., D. Pearce, and E. Stacchetti. 1986. Optimal Cartel Equilibria with Imperfect Monitoring. *Journal of Economic Theory* 39:251–269.

Aghion, P., and B. Hermalin. 1990. Legal Restrictions on Private Contracts Can Increase Efficiency. *Journal of Law, Economics, and Organization* 6:381–409.

Akerlof, G. 1970. The Market for Lemons: Quality Uncertainty and the Market Mechanism. *Quarterly Journal of Economics* 84:488–500.

Arrow, K. 1979. The Property Rights Doctrine and Demand Revelation under Incomplete Information. In *Economies and Human Welfare*. San Diego, Calif.: Academic Press.

d'Aspremont, C., and L.-A. Gerard-Varet. 1979. Incentives and Incomplete Information. *Journal of Public Economics* 11:25–45.

Ausubel, L., and R. Deneckere. 1989. A Direct Mechanism Characterization of Sequential Bargaining with One-Sided Incomplete Information. *Journal of Economic Theory* 48:18–46.

Axelrod, R. 1984a. *The Evolution of Cooperation*. New York: Basic Books.

———— 1984b. The Problem of Cooperation. In T. Cowen, ed., *The Theory of Market Failure: A Critical Examination*. Fairfax, Va.: George Mason University Press.

Ayres, I. 1990. Playing Games with the Law. *Stanford Law Review* 42:1291–1317.

Ayres, I., and R. Gertner. 1989. Filling Gaps in Incomplete Contracts: An Economic Theory of Default Rules. *Yale Law Journal* 99:87–130.

———— 1992. Strategic Contractual Inefficiency and the Optimal Choice of Legal Rules. *Yale Law Journal* 101:729–773.

Babcock, L., G. Loewenstein, S. Issacharoff, and C. Camerer. 1992. Biased Judgments of Fairness in Bargaining. Working Paper 92–15. School of Urban and Public Affairs, Carnegie Mellon University.

Baer, J., and O. Saxon. 1949. *Commodity Exchanges and Futures Trading*. New York: Harper and Brothers.

Baird, D., and T. Jackson. 1984. Corporate Reorganizations and the Treatment of Diverse Ownership Interests: A Comment on Adequate Protection of Secured Creditors in Bankruptcy. *University of Chicago Law Review* 51:97–130.

Baird, D., and R. Picker. 1991. A Simple Noncooperative Bargaining Model of Corporate Reorganizations. *Journal of Legal Studies* 20:311–349.

Banerjee, A. 1992. A Simple Model of Herd Behavior. *Quarterly Journal of Economics* 107:797–817.

Banks, J., and J. Sobel. 1987. Equilibrium Selection in Signalling Games. *Econometrica* 55:647–662.

Bebchuk, L. 1984. Litigation and Settlement under Imperfect Information. *Rand Journal of Economics* 15:404–415.

———— 1993. Suits That Are Known to Be Made Solely to Extract a Settlement Offer. Mimeo. Harvard University Law School.

Bebchuk, L., and H. Chang. 1992. Bargaining and the Division of Value in Corporate Reorganization. *Journal of Law, Economics, and Organization* 8:253–279.

Bebchuk, L., and S. Shavell. 1991. Information and the Scope of Liability for Breach of Contract: The Rule of *Hadley v. Baxendale. Journal of Law, Economics, and Organization* 7:284–312.

Becker, G. 1971. *Economic Theory.* New York: Alfred E. Knopf, Inc.

———— 1993. Nobel Lecture: The Economic Way of Looking at Life. *Journal of Political Economy* 101:385–409.

Ben-Porath, E., and E. Dekel. 1992. Signaling Future Actions and the Potential for Sacrifice. *Journal of Economic Theory* 57:36–51.

Bergman, Y., and J. Callen. 1991. Opportunistic Underinvestment in Debt Renegotiations and Capital Structure. *The Journal of Financial Economics* 29:137–171.

Bergström, C., P. Högfeldt, and K. Lithell. 1993. Restructuring of Financially Distressed Firms: The Role of Large Debtholders in Continental European and Scandinavian Financial Markets. Mimeo. Stockholm School of Economics.

Binmore, K. 1992. *Fun and Games: A Text on Game Theory.* Lexington, Mass.: D. C. Heath and Co.

Bolton, P., and J. Farrell. 1990. Decentralizaion, Duplication, and Delay. *Journal of Political Economy* 98:803–826.

Brilmayer, L. 1991. *Conflict of Laws: Foundations and Future Directions.* Boston: Little, Brown and Co.

Brown, J. 1973. Toward an Economic Theory of Liability. *Journal of Legal Studies* 2:323–349.

Buchanan, J. 1989. The Coase Theorem and the Theory of the State. In R. Tollison and V. Vanberg, eds., *Explorations into Constitutional Economics.* College Station, Tex.: Texas A & M University Press.

Calabresi, G. 1970. *The Costs of Accidents: A Legal and Economic Analysis.* New Haven, Conn.: Yale University Press.

———— 1975. Optimal Deterrence and Accidents. *Yale Law Journal* 84:656–671.

Calabresi, G., and J. Hirschoff. 1972. Toward a Test for Strict Liability in Torts. *Yale Law Journal* 81:1055–1085.

Carlton, D., and J. Klamer. 1983. The Need for Coordination among Firms with Special Reference to Network Industries. *University of Chicago Law Review* 50:446–465.

Carlton, D., and J. Perloff. 1990. *Modern Industrial Organization*. Glenview, Ill.: Scott Foresman / Little, Brown and Company.

Chatterjee, K., and L. Samuelson. 1987. Bargaining with Two-Sided Incomplete Information: An Infinite Horizon Model with Alternating Offers. *Review of Economic Studies* 54:175–192.

——— 1988. Bargaining under Two-Sided Incomplete Information: The Unrestricted Offers Case. *Operations Research* 36:605–618.

Chatterjee, K., and W. Samuelson. 1983. Bargaining under Incomplete Information. *Operations Research* 31:835–851.

Cho, I.-K., and D. Kreps. 1987. Signalling Games and Stable Equilibria. *Quarterly Journal of Economics* 102:179–221.

Chung, T.-Y. 1992. Efficiency of Comparative Negligence: A Game Theoretic Analysis. Mimeo. University of Western Ontario.

Clarke, E. 1971. Multipart Pricing of Public Goods. *Public Choice* 11:17–31.

Coase, R. 1937. The Nature of the Firm. *Economica* 4:386–405.

——— 1960. The Problem of Social Cost. *Journal of Law and Economics* 3:1–44.

——— 1972. Durability and Monopoly. *Journal of Law and Economics* 15:143–149.

——— 1988. *The Firm, the Market, and the Law*. Chicago: University of Chicago Press.

Cooper, R., D. DeJong, R. Forsythe, and T. Ross. 1990. Selection Criteria in Coordination Games: Some Experimental Results. *American Economic Review* 80:218–233.

Cooter, R. 1987. Why Litigants Disagree: A Comment on George Priest's "Measuring Legal Change." *Journal of Law, Economics, and Organization* 3:227–241.

Cooter, R., S. Marks, and R. Mnookin. 1982. Bargaining in the Shadow of the Law: A Testable Model of Strategic Behavior. *Journal of Legal Studies* 11:225–251.

Cooter, R., and T. Ulen. 1986. An Economic Case for Comparative Negligence. *New York University Law Review* 61:1067–1110.

——— 1988. *Law and Economics*. Glenview, Ill.: Scott, Foresman and Company.

Crocker, K., and S. Masten. 1991. Pretia Ex Machina? Prices and Process in Long-Term Contracts. *Journal of Law and Economics* 34:69–99.

Curran, C. 1992. The Spread of the Comparative Negligence Rule in the United States. *International Review of Law and Economics* 12:317–332.

Daughety, A., and J. Reinganum. 1991. Endogenous Sequencing in Models of Settlement and Litigation. Unpublished manuscript. University of Iowa.

Deaton, A., and J. Muellbauer. 1990. *Economics and Consumer Behavior*. Cambridge, England: Cambridge University Press.

Dewatripont, M. 1988. Commitment through Renegotiation-Proof Contracts with Third Parties. *Review of Economic Studies* 55:377–390.

Diamond, P. 1974. Accident Law and Resource Allocation. *Bell Journal of Economics and Management Science* 5:366–405.

Dixit, A. 1979. A Model of Duopoly Suggesting a Theory of Entry Barriers. *Bell Journal of Economics* 10:20–32.

Dixit, A., and B. Nalebuff. 1991. *Thinking Strategically: The Competitive Edge in Business, Politics, and Everyday Life.* New York: Norton.

Dye, R. 1985. Costly Contract Contingencies. *International Economic Review* 26:233–250.

Easterbrook, F. 1981. Predatory Strategies and Counterstrategies. *University of Chicago Law Review* 48:263–337.

——— 1989. Comment: Discovery as Abuse. *Boston University Law Review* 69:635–648.

Easterbrook, F., and D. Fischel. 1991. *The Economic Structure of Corporate Law.* Cambridge, Mass.: Harvard University Press.

Eisenberg, T. 1990. Testing the Selection Effect: A New Theoretical Framework with Empirical Tests. *Journal of Legal Studies* 19:337–358.

Ellickson, R. 1991. *Order without Law: How Neighbors Settle Disputes.* Cambridge, Mass.: Harvard University Press.

Elster, J. 1986. *Rational Choice.* New York: New York University Press.

——— 1989. *The Cement of Society: A Study of Social Order.* Cambridge, England: Cambridge University Press.

Emons, W. 1993. The Provision of Environmental Protection Measures under Incomplete Information: An Introduction to the Theory of Mechanism Design. Mimeo. University of Bern.

Epstein, R. 1989. Beyond Foreseeability: Consequential Damages in the Law of Contract. *Journal of Legal Studies* 18:105–138.

Farber, D., and P. Frickey. 1991. *Law and Public Choice: A Critical Introduction.* Chicago: University of Chicago Press.

Farrell, J. 1983. Communication in Games I: Mechanism Design without a Mediator. MIT Working Paper.

——— 1987. Information and the Coase Theorem. *Journal of Economic Perspectives* 1:113–129.

——— 1989. Standardization and Intellectual Property. *Jurimetrics Journal* 30:35–50.

Farrell, J., and R. Gibbons. 1989. Cheap Talk Can Matter in Bargaining. *Journal of Economic Theory* 48:221–237.

Farrell, J., and E. Maskin. 1989. Renegotiation in Repeated Games. *Games and Economic Behavior* 1:327–360.

Farrell, J., and G. Saloner. 1985. Standardization, Compatibility, and Innovation. *Rand Journal of Economics* 16:70–83.

——— 1986. Installed Base and Compatibility: Innovation, Product Preannouncements, and Predation. *American Economic Review* 76:940–955.

Fernandez, R., and J. Glazer. 1991. Striking for a Bargain between Two Completely Informed Agents. *American Economic Review* 81:240–252.

Fishman, M., and K. Hagerty. 1990. The Optimal Amount of Discretion to Allow in Disclosure. *Quarterly Journal of Economics* 105:427–444.

Friedman, D. 1989. An Economic Analysis of Alternative Damage Rules for Breach of Contract. *The Journal of Law and Economics* 32:281–310.

——— 1992. Deterring Imperfectly Informed Tortfeasors: Optimal Rules for Penalty and Liability. Mimeo. University of Chicago.

Friedman, J. 1990. *Game Theory with Applications to Economics*. 2nd edition. Oxford: Oxford University Press.

Fudenberg, D., and J. Tirole. 1991a. *Game Theory*. Cambridge, Mass.: MIT Press.

——— 1991b. Perfect Bayesian Equilibrium and Sequential Equilibrium. *Journal of Economic Theory* 53:236–260.

Geanakoplos, J. 1992. Common Knowledge. *Journal of Economic Perspectives* 6, no. 4: 53–82.

Ghemawat, P. 1984. Capacity Expansion in the Titanium Dioxide Industry. *Journal of Industrial Economics* 33:145–163.

Gibbons, R. 1988. Learning in Equilibrium Models of Arbitration. *American Economic Review* 78:896–912.

——— 1992. *Game Theory for Applied Economists*. Princeton, N.J.: Princeton University Press.

Goetz, C., and R. Scott. 1985. The Limits of Expanded Choice: An Analysis of the Interactions between Express and Implied Contract Terms. *California Law Review* 73:261–322.

Gordon, J. 1991. Shareholder Initiative: A Social Choice and Game Theoretic Approach to Corporate Law. *University of Cincinnati Law Review* 60: 347–385.

Gould, J. 1973. The Economics of Legal Conflicts. *Journal of Legal Studies* 2:279–300.

Green, E., and R. Porter. 1984. Non-cooperative Collusion under Imperfect Price Information. *Econometrica* 52:87–100.

Grossman, S. 1981. The Informational Role of Warranties and Private Disclosure about Product Quality. *Journal of Law and Economics* 24:461–484.

Grossman, S., and O. Hart. 1980. Disclosure Laws and Takeover Bids. *Journal of Finance* 35:323–334.

——— 1986. The Costs and Benefits of Ownership: A Theory of Vertical and Lateral Integration. *Journal of Political Economy* 94:691–719.

Grossman, S., and M. Perry. 1986. Perfect Sequential Equilibrium. *Journal of Economic Theory* 39:97–119.

Groves, T. 1973. Incentives in Teams. *Econometrica*, 41:617–631.

Gul, F., and H. Sonnenschein. 1988. On Delay in Bargaining with One-Sided Uncertainty. *Econometrica* 56:601–11.

Gul, F., H. Sonnenschein, and R. Wilson. 1986. Foundations of Dynamic Monopoly and the Coase Conjecture. *Journal of Economic Theory* 39:155–190.

Hadfield, G. 1990. Problematic Relations: Franchising and the Law of Incomplete Contracts. *Stanford Law Review* 42:927–992.

Hammitt, J., S. Carroll, and D. Relles. 1985. Tort Standards and Jury Decisions. *Journal of Legal Studies* 14:751–762.

Hampton, J. 1992. Expected Utility Theory. Unpublished manuscript.

Hardin, G. 1968. The Tragedy of the Commons. *Science* 162:1243–1248.

Hardin, R. 1982. *Collective Action*. Baltimore: Johns Hopkins University Press.

Hart, O., and B. Holmstrom. 1987. The Theory of Contracts. In Truman Bewley, ed., *Advances in Economic Theory*. Cambridge, England: Cambridge University Press.

Hart, O., and J. Moore. 1988. Incomplete Contracts and Renegotiation. *Econometrica* 56:755–785.

Hirshleifer, J., and J. Riley. 1992. *The Analytics of Uncertainty and Information*. Cambridge, England: Cambridge University Press.

Hughes, J., and E. Snyder. 1991. Litigation under the English and American Rules: Theory and Evidence. Unpublished manuscript. University of Michigan.

Hylton, K. 1990. Costly Litigation and Legal Error under Negligence. *Journal of Law, Economics, and Organization* 6:433–452.

——— 1992. Litigation Cost Allocation Rules and Compliance with the Negligence Standard. Mimeo. Northwestern School of Law.

Jackson, T. 1982. Bankruptcy, Non-Bankruptcy Entitlements, and the Creditors' Bargain. *Yale Law Journal* 91:857–907.

——— 1986. *The Logic and Limits of Bankruptcy Law*. Cambridge, Mass.: Harvard University Press.

Johnston, J. 1990. Strategic Bargaining and the Economic Theory of Contract Default Rules. *Yale Law Journal* 100:615–664.

Joskow, P. 1985. Vertical Integration and Long-Term Contracts: The Case of Coal-Burning Electric Generating Plants. *Journal of Law, Economics, and Organization* 1:33–80.

Jovanovic, B. 1982. Truthful Disclosure of Information. *Bell Journal of Economics* 13:36–44.

Kahneman, D., J. Knetsch, and R. Thaler. 1990. Experimental Tests of the Endowment Effect and the Coase Theorem. *Journal of Political Economy* 98:1325–1349.

——— 1991. The Endowment Effect, Loss Aversion, and Status Quo Bias: Anomalies. *Journal of Economic Perspectives* 5:193–206.

Kaplow, L. 1992. Rules versus Standards: An Economic Analysis. *Duke Law Journal* 42:557–629.

Katz, A. 1990a. The Effect of Frivolous Lawsuits on the Settlement of Litigation. *International Review of Law and Economics* 10:3–27.

——— 1990b. The Strategic Structure of Offer and Acceptance: Game Theory and the Law of Contract Formation. *Michigan Law Review* 89:216–295.

—— 1990c. Your Terms or Mine? The Duty to Read the Fine Print in Contracts. *Rand Journal of Economics* 21:518–537.

Katz, L. 1987. *Bad Acts and Guilty Minds: Conundrums of the Criminal Law.* Chicago: University of Chicago Press.

Katz, M., and C. Shapiro. 1985. Network Externalities, Competition, and Compatibility. *American Economic Review* 75:424–440.

—— 1986. Technology Adoption in the Presence of Network Externalities. *Journal of Political Economy* 94:822–841.

Kennan, J., and R. Wilson. 1989. Strategic Bargaining Models and Interpretation of Strike Data. *Journal of Applied Econometrics* 4:S87-S130.

—— 1990. Can Strategic Bargaining Models Explain Collective Bargaining Data? *American Economic Review: Proceedings* 80:405–409.

Klein, B., R. Crawford, and A. Alchian. 1978. Vertical Integration, Appropriable Rents, and the Competitive Contracting Process. *Journal of Law and Economics* 21:297–326.

Kohlberg, E. 1990. Refinement of Nash Equilibrium: The Main Ideas. In T. Ichiishi, A. Neyman, and Y. Tauman, eds., *Game Theory and Applications.* San Diego, Calif.: Academic Press.

Kohlberg, E., and J.-F. Mertens. 1986. On the Strategic Stability of Equilibria. *Econometrica* 54:1003–1038.

Kornhauser, L., and R. Revesz. 1991. Sequential Decisions by a Single Tortfeasor. *Journal of Legal Studies* 20:363–380.

Kramer, L. 1990. Rethinking Choice of Law. *Columbia Law Review* 90:277–345.

Kreps, D. 1988. *Notes on the Theory of Choice.* London: Westview Press.

—— 1990a. *Game Theory and Economic Modeling.* Oxford: Clarendon Press.

—— 1990b. *A Course in Microeconomic Theory.* Princeton, N.J.: Princeton University Press.

Kreps, D., and R. Wilson. 1982. Reputation and Imperfect Information. *Journal of Economic Theory* 27:253–279.

Kreps, D., P. Milgrom, J. Roberts, and R. Wilson. 1982. Rational Cooperation in the Finitely Repeated Prisoners' Dilemma. *Journal of Economic Theory* 27:245–252.

Kronman, A. 1978. Mistake, Disclosure, Information, and the Law of Contracts. *Journal of Legal Studies* 7:1–34.

—— 1985. Contract Law and the State of Nature. *The Journal of Law, Economics, and Organization* 1:5–32.

Kydland, F., and E. Prescott. 1977. Rules Rather Than Discretion: The Inconsistency of Optimal Plans. *Journal of Political Economy* 85:473–492.

Laffont, J.-J. 1988. *Fundamentals of Public Economics.* Cambridge, Mass.: MIT Press.

—— 1990. *The Economics of Uncertainty and Information.* Cambridge, Mass.: MIT Press.

Laffont, J.-J., and E. Maskin. 1982. The Theory of Incentives: An Overview. In

W. Hildenbrand, ed., *Advances in Economic Theory.* Cambridge, England: Cambridge University Press.

Laffont, J.-J., and J. Tirole. 1993. Commitment and Renegotiation. In *A Theory of Incentives in Procurement and Regulation* Cambridge, Mass.: MIT Press.

Landes, W. 1971. An Economic Analysis of the Courts. *Journal of Law and Economics* 14:61–107.

———— 1993. Sequential versus Unitary Trials: An Economic Analysis. *Journal of Legal Studies* 22:99–134.

Landes, W., and R. Posner. 1980. Joint and Multiple Tortfeasors: An Economic Analysis. *Journal of Legal Studies* 9:517–556.

———— 1981. The Positive Economic Theory of Tort Law. *Georgia Law Review* 15:851–924.

———— 1987. *The Economic Structure of Tort Law.* Cambridge, Mass.: Harvard University Press.

Leebron, D. 1991. A Game Theoretic Approach to the Regulation of Foreign Direct Investment and the Multinational Corporation. *University of Cincinnati Law Review* 60:305–346.

Leinfellner, W. 1986. The Prisoner's Dilemma and Its Evolutionary Iteration. In A. Diekmann and P. Mitter, eds., *Paradoxical Effects of Social Behavior: Essays in Honor of Anatol Rapoport.* Vienna: Physica.

Loewenstein, G., S. Issacharoff, C. Camerer, and L. Babcock. 1993. Self-Serving Assessments of Fairness and Pretrial Bargaining. *Journal of Legal Studies* 22:135–60.

Lott, J., and T. Opler. 1992. Testing Whether Predatory Commitments are Credible. Unpublished manuscript. University of Pennsylvania.

Matsui, A. 1991. Cheap Talk and Cooperation in a Society. *Journal of Economic Theory* 54:245–258.

Maute, J. 1993. *Peevyhouse v. Garland Coal Co.* Revisited: The Ballad of Willie and Lucille. Unpublished manuscript. The University of Oklahoma.

McMillan, J. 1992. *Games, Strategies, and Managers.* Oxford: Oxford University Press.

Menell, P. 1987. Tailoring Legal Protection for Computer Software. *Stanford Law Review* 39:1329–72.

———— 1989. An Analysis of the Scope of Copyright Protection for Application Programs. *Stanford Law Review* 41:1045–1104.

Milgrom, P. 1981. Good News and Bad News: Representation Theorems and Applications. *Bell Journal of Economics* 12:380–391.

Milgrom, P., and J. Roberts. 1982. Predation, Reputation, and Entry Deterrence. *Journal of Economic Theory* 27:280–312.

Mnookin, R., and L. Kornhauser. 1979. Bargaining in the Shadow of the Law: The Case of Divorce. *Yale Law Journal* 88:950–997.

Mnookin, R., and R. Wilson. 1989. Rational Bargaining and Market Efficiency: Understanding *Pennzoil v. Texaco. Virginia Law Review* 75:295–334.

Myerson, R. 1979. Incentive Compatibility and the Bargaining Problem. *Econometrica* 47:61–73.

――― 1981. Optimal Auction Design. *Mathematics of Operations Research* 6:58–73.

――― 1991. *Game Theory: Analysis of Conflict*. Cambridge, Mass.: Harvard University Press.

Myerson, R., and M. Satterthwaite. 1983. Efficient Mechanisms for Bilateral Trading. *Journal of Economic Theory* 29:265–281.

Nalebuff, B. 1987. Credible Pretrial Negotiation. *Rand Journal of Economics* 18:198–210.

Nash, J. 1950a. The Bargaining Problem. *Econometrica* 18:155–162.

――― 1950b. Equilibrium Points in N-Person Games. *Proceedings of the National Academy of Sciences* (USA) 36:48–49.

Neale, M., and M. Bazerman. 1991. *Cognition and Rationality in Negotiation*. New York: The Free Press.

Nozick, R. 1985. Newcomb's Problem and Two Principles of Choice. In R. Campbell and L. Sowden, eds., *Paradoxes of Rationality and Cooperation: Prisoner's Dilemma and Newcomb's Problem*. Vancouver: University of British Columbia Press.

Okuno-Fujiwara, M., A. Postlewaite, and K. Suzumura. 1990. Strategic Information Revelation. *Review of Economic Studies* 57:25–47.

Olson, M. 1965. *The Logic of Collective Action: Public Goods and the Theory of Groups*. Cambridge, Mass.: Harvard University Press.

Ordover, J. 1978. Costly Litigation in the Model of Single Activity Accidents. *Journal of Legal Studies* 7:243–262.

――― 1981. On the Consequences of Costly Litigation in the Model of Single Activity Accidents: Some New Results. *Journal of Legal Studies* 10:269–291.

Ordover, J., and G. Saloner. 1989. Predation, Monopolization, and Antitrust. In R. Schmalensee and R. Willig, eds., *Handbook of Industrial Organization*. Volume 1. New York: North-Holland.

Orr, D. 1991. The Superiority of Comparative Negligence: Another Vote. *Journal of Legal Studies* 20:119–129.

Osborne, M., and A. Rubinstein. 1990. *Bargaining and Markets*. San Diego, Calif.: Academic Press.

Ostrom, E. 1990. *Governing the Commons: The Evolution of Institutions for Collective Action*. Cambridge, England: Cambridge University Press.

Palfrey, T., and H. Rosenthal. 1991. Testing for the Effects of Cheap Talk in a Public Goods Game with Private Information. *Games and Economic Behavior* 3:183–220.

Pearce, D. 1988. Renegotiation-Proof Equilibria: Collective Rationality and Intertemporal Cooperation. Unpublished manuscript. Yale University.

Picker, R. 1992. Security Interests, Misbehavior, and Common Pools. *University of Chicago Law Review* 59:645–679.

P'ng, I. 1983. Strategic Behavior in Suit, Settlement, and Trial. *Bell Journal of Economics* 14:539–550.

Polinsky, M. 1987. Optimal Liability When the Injurer's Information about the Victim's Loss Is Imperfect. *The International Review of Law and Economics* 7:139–147.

Posner, R. 1973. An Economic Approach to Legal Procedure and Judicial Administration. *Journal of Legal Studies* 2:399–458.

——— 1992. *Economic Analysis of Law.* 4th edition. Boston: Little, Brown and Co.

Poundstone, W. 1992. *Prisoner's Dilemma.* New York: Doubleday.

Priest, G. 1985. Reexamining the Selection Hypothesis: Learning from Wittman's Mistakes. *Journal of Legal Studies* 14:215–243.

Priest, G. 1987. Measuring Legal Change. *Journal of Law, Economics, and Organization* 3:193–225.

Priest, G., and B. Klein. 1984. The Selection of Disputes for Litigation. *Journal of Legal Studies* 13:1–55.

Rabin, M. 1990. Communication between Rational Agents. *Journal of Economic Theory* 51:144–170.

Raiffa, H. 1982. *The Art and Science of Negotiation.* Cambridge, Mass.: Harvard University Press.

Rasmusen, E. 1989. *Games and Information: An Introduction to Game Theory.* New York: Basil Blackwell.

Rea, S. 1987. The Economics of Comparative Negligence. *International Review of Law and Economics* 7:149–162.

Reinganum, J., and L. Wilde. 1986. Settlement, Litigation, and the Allocation of Litigation Costs. *Rand Journal of Economics* 17:557–566.

Roe, M. 1987. The Voting Prohibition in Bond Workouts. *Yale Law Journal* 97:232–279.

Rogerson, W. 1984. Efficient Reliance and Damage Measures for Breach of Contract. *Bell Journal of Economics* 15:39–53.

Rotemberg, J., and G. Saloner. 1986. A Supergame-Theoretic Model of Price Wars during Booms. *American Economic Review* 76:390–407.

Rothschild, M., and J. Stiglitz. 1976. Equilibrium in Competitive Insurance Markets: An Essay on the Economics of Imperfect Information. *Quarterly Journal of Economics* 90:629–650.

Rubinfeld, D. 1987. The Efficiency of Comparative Negligence. *Journal of Legal Studies* 16:375–394.

Rubinstein, A. 1982. Perfect Equilibrium in a Bargaining Model. *Econometrica* 50:97–110.

——— 1985. A Bargaining Model with Incomplete Information about Time Preferences. *Econometrica* 53:1151–1172.

Sarig, O. 1988. Bargaining with a Corporation and the Capital Structure of the Bargaining Firm. Unpublished manuscript. Tel Aviv University.

Schelling, T. 1960. *The Strategy of Conflict.* Cambridge, Mass.: Harvard University Press.

Scherer, F., and D. Ross. 1990. *Industrial Market Structure and Economic Performance.* 3rd edition. Boston: Houghton Mifflin.

Schmalensee, R., and R. Willig, eds. 1989. *Handbook of Industrial Organization.* Volumes 1 and 2. New York: North-Holland.

Schwartz, A. 1992. Relational Contract in the Courts: An Analysis of Incomplete Agreements and Judicial Strategies. *Journal of Legal Studies* 21:271–318.

Schwartz, G. 1978. Contributory and Comparative Negligence: A Reappraisal. *Yale Law Journal* 87:697–727.

Schweizer, U. 1989. Litigation and Settlement under Two-Sided Incomplete Information. *Review of Economic Studies* 56:163–177.

Seidmann, D. 1990. Effective Cheap Talk with Conflicting Interests. *Journal of Economic Theory* 50:445–458.

Selten, R. 1975. Reexamination of the Perfectness Concept for Equilibrium Points in Extensive Games. *International Journal of Game Theory* 4:25–55.

—— 1978. The Chain-Store Paradox. *Theory and Decision* 9:127–159.

Setear, J. 1989. The Barrister and the Bomb: The Dynamics of Cooperation, Nuclear Deterrence, and Discovery Abuse. *Boston University Law Review* 69:569–633.

Shavell, S. 1980. Damage Measures for Breach of Contract. *Bell Journal of Economics* 11:466–490.

—— 1987. *Economic Analysis of Accident Law.* Cambridge, Mass.: Harvard University Press.

—— 1989. Sharing of Information Prior to Settlement or Litigation. *Rand Journal of Economics* 20:183–195.

—— 1991. Acquisition and Disclosure of Information Prior to Economic Exchange. Unpublished manuscript. Harvard University Law School.

Snyder, E., and J. Hughes. 1990. The English Rule for Allocating Legal Costs: Evidence Confronts Theory. *Journal of Law, Economics, and Organization* 6:345–380.

Sobel, J. 1989. An Analysis of Discovery Rules. *Law and Contemporary Problems* 52:133–159.

Sobelsohn, D. 1985. Comparing Fault. *Indiana Law Journal* 60:413–462.

Spence, M. 1974. *Market Signalling: Informational Transfer in Hiring and Related Screening Processes.* Cambridge, Mass.: Harvard University Press.

—— 1977. Entry, Capacity, Investment, and Oligopolistic Pricing. *Bell Journal of Economics* 8:534–544.

Spier, K. 1988. Incomplete Contracts in a Model with Adverse Selection and Exogenous Costs of Enforcement. Unpublished manuscript. Harvard University.

—— 1989. Efficient Mechanisms for Pretrial Bargaining. In "Three Essays on Dispute Resolution and Incomplete Contracts." Ph.D. dissertation. MIT.

——— 1992. The Dynamics of Pretrial Negotiation. *Review of Economic Studies* 59:93–108.

Spulber, D. 1990. Contingent Damages and Settlement Bargaining. Unpublished manuscript. J. L. Kellogg Graduate School of Management.

Stigler, G. 1964. A Theory of Oligopoly. *Journal of Political Economy* 72:44–61.

Stole, L. 1992. The Economics of Liquidated Damage Clauses in Contractual Environments with Private Information. *Journal of Law, Economics and Organization* 8:582–606.

Sunstein, C. 1990. *After the Rights Revolution: Reconceiving the Regulatory State.* Cambridge, Mass.: Harvard University Press.

——— 1993. Endogenous Preferences, Environmental Law. University of Chicago Law School Working Paper no. 14.

Sutton, J. 1986. Noncooperative Bargaining Theory: An Introduction. *Review of Economic Studies* 53:709–724.

Thaler, R. 1991. *Quasi-Rational Economics.* New York: Russell Sage Foundation.

Tietenberg, T. 1989. Indivisible Toxic Torts: The Economics of Joint and Several Liability. *Land-Economics* 65:305–319.

Tirole, J. 1988. *The Theory of Industrial Organization.* Cambridge, Mass.: MIT Press.

Tsebelis, G. 1990. *Nested Games: Rational Choice in Comparative Politics.* Berkeley: University of California Press.

van Damme, E. 1989. Stable Equilibria and Forward Induction. *Journal of Economic Theory* 48:476–496.

Varian, H. 1978. *Microeconomic Analysis.* New York: Norton.

——— 1992. *Microeconomic Analysis.* 3rd edition. New York: Norton.

Vickrey, W. 1961. Counterspeculation, Auctions, and Competitive Sealed Tenders. *Journal of Finance* 16:8–37.

Vincent, D. 1989. Bargaining with Common Values. *Journal of Economic Theory* 48:47–62.

von Neumann, J. 1928. Zur Theorie der Gesellschaftsspiele. *Mathematische Annalen* 100:295–320.

von Neumann, J., and O. Morgenstern. 1944. *Theory of Games and Economic Behavior.* Princeton, N.J.: Princeton University Press.

White, M. 1989. An Empirical Test of the Comparative and Contributory Negligence Rules in Accident Law. *Rand Journal of Economics* 20:308–330.

Wiley, J. 1988. Reciprocal Altruism as a Felony: Antitrust and the Prisoner's Dilemma. *Michigan Law Review* 86:1906–1928.

Williamson, O. 1983. Credible Commitments: Using Hostages to Support Exchange. *American Economic Review* 73:519–540.

——— 1985. *The Economic Institutions of Capitalism.* New York: The Free Press.

Wittman, D. 1985. Is the Selection of Cases for Trial Biased? *Journal of Legal Studies* 14:185–214.

——— 1988. Dispute Resolution, Bargaining, and the Selection of Cases for Trial: A Study of the Generation of Biased and Unbiased Data. *Journal of Legal Studies* 17:313–352.

Glossary

Absolute priority rule. See *new value exception*

Adverse selection. A problem of hidden information. Parties with private, *non-verifiable information* know whether they will impose additional costs on their contracting opposites. Those parties who impose the highest costs will be disproportionately likely to enter into a contract at a given price. Because of this, the price of such contracts rises, and the effect is magnified further. Problems of adverse selection arise in insurance and in many other contracting environments.

American rule. A rule of civil procedure that provides that both winning and losing litigants bear their own attorney's fees.

Assurance game. Closely related to the *stag hunt game*. Both parties prefer to cooperate with each other and engage in a common enterprise (such as hunting stag). If one player decides not to hunt stag, however, that player is better off if the other player does hunt stag. (Hunting hare may have a *payoff* of $8 when the other hunts stag, but only $6 when the other player hunts hare as well.) In the assurance game, as in the *stag hunt game,* there are two *pure strategy Nash equilibria* as well as a *mixed strategy Nash equilibrium*.

Automatic stay. Outside of bankruptcy, creditors are typically allowed to seize property of the debtor in the event of default. The automatic stay is a rule of bankruptcy law that prevents a creditor from exercising this right during the pendency of the bankruptcy without court permission. A court will lift the automatic stay and allow the creditor to reach *collateral* for cause; it will also lift the automatic stay when the debtor has no equity in the property and the property is not necessary for an effective reorganization.

Axiomatic approach. See *cooperative bargaining*

Axioms. In the *axiomatic approach* to bargaining, the characteristics that one posits a bargain is likely to have, such as that it is likely to be *Pareto-optimal*.

Backwards induction. A *solution concept* applicable to an *extensive form game* made up of *information sets* consisting of single *nodes*. To use it, one examines the actions available to a player at the *decision nodes* immediately preceding the *terminal nodes*. One identifies the action that brings that player the highest *payoff* and then writes a new *extensive form game* in which the *decision node* is replaced with the *terminal node* associated with that *payoff*. One then repeats the process until one is left with only a single *terminal node*. This *terminal node* is the outcome of the game.

Bargained-for share. The amount a player receives in a simple *Rubinstein bargaining game* in which there are no *exit options*.

Battle-of-the-sexes game. A paradigmatic, two-by-two *normal form* coordination game. There is a conflict between two people who want to spend the evening together but have different preferences about whether to go to a fight or to an opera. Both would rather be with the other at the event they liked least than go alone to the event they liked most, but both would like most for the other to be with them at their favored event. Neither, however, is able to communicate with the other. Each must guess what the other will do. It is a *Nash equilibrium* for both to go to the fight, for both to go to the opera, or for each to randomize between the two. This game exemplifies coordination games in which there are multiple *Nash equilibria* but individual players have a preference for one rather than another. Both players want to coordinate their actions, but each player wants a different outcome.

Bayes's rule. A rule that provides a method for updating *beliefs* in light of new information. For example, we may want to know what *belief* a player should have about another player's type after the former has observed that the latter has taken some action. Bayes's rule tells us how a player must adjust that player's prior *belief* by taking into account both the likelihood that the player will observe this action and the likelihood that a player of a certain type will take this action. This rule is the rational way to update *beliefs*. Bayes's rule, for example, would require that the new *belief* about a player's type be the same as the old one in a proposed *equilibrium* in which all players take the same action. In such an *equilibrium*, the actions of the players have provided no new information. Similarly, Bayes's rule requires the uninformed player to believe that anyone who takes an action is of a particular type if, in a proposed *equilibrium*, players of a particular type—and only those players—take a particular action. For a formal definition and derivation of Bayes's rule, see Rasmusen (1989).

Beer-quiche game. A *signaling* game invented by Cho and Kreps (1987) to explore the *equilibrium dominance* refinement to the *perfect Bayesian equilibrium*. In this game, one player is a bully and the other player is a patron of a restaurant. The patron is one of two types, a tough guy or a wimp. The patron must decide what to order for breakfast (beer or quiche), and the bully must decide, after seeing what the patron orders, whether to pick a fight. Neither tough guys nor wimps like to fight. The former prefers beer for breakfast, the latter quiche. Bullies like to fight wimps, but they do not like to fight tough guys.

Beliefs. When a player must move in an *extensive form game* and is at an *information set* that contains multiple *nodes*, the player has beliefs about the probability of being at each of the *nodes*. The *perfect Bayesian equilibrium solution concept* requires that a player's beliefs be consistent with *Bayes's rule*.

Bimatrix. A matrix in which there are two numbers in each cell. It is the standard way of illustrating a *normal form game* in which there are two players, each of whom has a small number of *strategies*. By convention, the first value in a bimatrix is the *payoff* associated with the row player and the second is the value associated with the column player.

Boulwareism. A labor negotiation policy on the part of management, named after Lemuel Boulware, a vice president of General Electric. Management purports to make a "fair and reasonable offer" at the start of negotiations that it will change only if it has overlooked some issue. Otherwise, management claims that it will neither budge from the first offer nor engage in the usual practice of splitting the difference. The legal question courts face is whether the practice of starting a round of collective bargaining with a final "fair, firm offer" is a breach of the duty to bargain in good faith. There is not a per se prohibition against the practice, but "an employer may not so combine 'take-it-or-leave-it' bargaining methods with a widely publicized stance of unbending firmness that he is himself unable to alter a position once taken."

Branch. The fundamental building block of the *extensive form game*. It is an action available to a player at a particular *node*.

Chain-store paradox. When we solve *repeated games* of finite length using *backwards induction*, the outcome in every game is the same as it would be if the game were played in isolation. This idea that *backwards induction* leads to unraveling back to the first move is called the chain-store paradox because it was developed in the context of *predatory pricing* with an incumbent firm that owned a chain of stores in many towns and faced a potential entrant in each town.

Cheap talk. When a player makes a statement that may convey information even though the statement is costless, nonbinding, and nonverifiable.

Chicken game. A two-by-two *normal form game* that captures the following interaction: Two teenagers drive cars headlong at each other. A driver gains stature when that driver drives headlong and the other swerves. Both drivers die, however, if neither swerves. Each player's highest *payoff* comes when that player drives head on and the other swerves. The second highest payoff comes when that player swerves and the other player swerves as well. The third high-est comes when that player swerves and the other drives. The lowest *payoff* is when both drive. This is a game of multiple *Nash equilibria,* but unlike the *stag hunt game,* the *assurance game,* or the *battle-of-the-sexes game,* the *pure strategy equilibria* are ones in which each player adopts a different action (that is, one swerves and the other drives).

Cho-Kreps refinements. A set of ways in which implausible equilibria in games such as *beer-quiche* are eliminated by positing what *beliefs* a rational player is likely to have about actions *off the equilibrium path.*

Civil damages. At common law, a successful plaintiff is ordinarily given the right to collect damages from the defendant. A rule that provides for a transfer of wealth from one party to another is a rule of civil damages, as distinguished from alternative regimes, such as ones in which part of the damage award goes to the state or in which the defendant is compelled to perform a particular act. In games of *complete but imperfect information,* when a unique social optimum exists, civil damages can implement it in *dominant strategies.*

Clarke-Groves mechanism. A *direct mechanism* for eliciting private, nonverifi-able information in which truth-telling is a *weakly dominant strategy* for all the players. Such a mechanism, however, cannot ensure a socially optimal outcome with a balanced budget. The amount that some players lose cannot be exactly offset by the amount that others gain.

Coase conjecture. A monopolist who sells a durable good will set a price that approaches the competitive price as the period between sales becomes arbi-trarily short.

Coase theorem. When information is complete, transaction costs are zero, and barriers to bargaining are nonexistent, private and social costs are equal.

Collateral. See *security interest*

Common knowledge. Something is common knowledge if it is known to each player, and, in addition, each player knows that the other player has this knowledge; knows that the other person knows the player knows it; and so forth.

Comparative negligence. Under a legal regime of comparative negligence, a victim who acts negligently is entitled to recover damages from an injurer, but the victim's recovery is reduced by taking account of the relative fault of the victim.

Complaint. A simple, written document a plaintiff files to begin a civil action. It explains why the court has jurisdiction, sets out the grounds on which the plaintiff relies for relief, and specifies the relief that the plaintiff seeks.

Complete information. In a game of complete information, the structure of the game and the *payoffs* to every player (but not necessarily the actions that other players take) are *common knowledge*. Information is incomplete if at least one player is uncertain about the structure of the game or the *payoffs* to another player.

Contributory negligence. The victim of an accident is contributorily negligent whenever the victim fails to exercise reasonable care. Under a regime of *negligence* coupled with contributory negligence, or *strict liability* coupled with contributory negligence, a victim is entitled to recover damages from an injurer only if the victim is not contributorily negligent.

Cooperative bargaining. This approach, also called the *axiomatic approach*, posits a series of *axioms* about the characteristics that agreement between rational players should have and examines the conditions under which such agreements are possible and whether the result is unique. The *Nash program* is an effort to connect the principles of cooperative bargaining with those of *noncooperative bargaining*.

Credible threat. A threat is credible when a player would find it in that player's self-interest to carry out the threat if ever called upon to do so. *Subgame perfection* and *backwards induction* are formal tools that ensure that solutions to games do not depend upon threats that are not credible.

Decision node. See *node*

Default rules. Most rules of contracts allocate rights between parties in the absence of any explicit mention of such terms in the contract. As such, these rules are simply presumptions. Because parties are free to opt out of them if they choose, they are often called default rules—rules that apply by default if the parties do not address the issue.

Direct mechanism. A game in which the only action of the players is to announce their type. This is unlike some other allocation mechanisms in which the players take some other action that might reveal their type. If any allocation mechanism leads to a particular *equilibrium*, there also exists a direct mechanism in which the players truthfully reveal their type that implements the same *equilibrium*. If no direct mechanism in which parties truthfully reveal their type exists that reaches a particular *equilibrium*, no other scheme that we can devise can reach it either. This theorem is called the *revelation principle*.

Discount factor. In a repeated game, it is natural to assume that a player values *payoffs* in future periods less than a *payoff* in the present period. A discount factor is the amount by which the value of a *payoff* in the next period must be adjusted to reflect its value in the present period. If we have a discount factor of δ, the present value of one dollar earned in the subsequent period is δ dollars. This implies that the interest rate is $(1 - \delta)/\delta$.

Discovery. The process by which civil litigants are entitled to demand relevant evidence from each other. Generally speaking, a civil litigant may obtain discovery regarding any matter, not privileged, that is relevant to the subject matter involved in the litigation. A discovery request must list the items to be inspected either by individual item or by category, and describe each item and category with reasonable particularity.

Dominant strategy. A *strategy* that is a best choice for a player in a game for every possible choice by the other player. When one *strategy* is no better than another *strategy*, and sometimes worse, it is dominated by that *strategy*. When one *strategy* is always worse than another, it is strictly dominated. In the text, we use dominant strategy as a shorthand for a strictly dominant strategy, unless otherwise noted. Dominant and *dominated strategies* are the first and perhaps most important *solution concepts* of *game theory*: A player will choose a strictly dominant strategy whenever possible and will not choose any strategy that is strictly dominated by another.

Dominated strategy. See *dominant strategy*

Due care. The amount of care that each player takes when the total social cost of an accident is minimized.

English rule. A rule of civil procedure that provides that losing litigants bear both their own attorney's fees and those of the prevailing party.

Equilibrium. A combination of *strategies* and *beliefs* that the players of a game are likely to adopt.

Equilibrium dominance. A *refinement* of the *perfect Bayesian solution concept*. It requires that a player's *beliefs* about behavior *off the equilibrium path* should be such that this player will believe that, if a deviation were to take place, the player who deviates would, if at all possible, not be of a type whose *payoff* in the proposed *equilibrium* is larger than any *payoff* that player could ever receive by deviating. This *refinement* is also called the *intuitive criterion* or the *Cho-Kreps refinement*. Just as a player should not expect another to play a *dominated strategy*, a player should not expect a player to choose an action that yields a lower *payoff* than that player receives in the proposed equilibrium.

Equilibrium path. In an *extensive form game*, the equilibrium path is the sequence of actions the players adopt with some positive probability in a proposed equilibrium. *Nodes* that are *off the equilibrium path* are *nodes* that are never actually reached in the proposed equilibrium, and hence the parties never actually take the actions at these *nodes* that are part of the proposed equilibrium. Nevertheless, the actions and *beliefs* of players at these *nodes* matter because they affect what other players do.

Exclusivity period. In a corporate reorganization under Chapter 11 of the Bankruptcy Code, the debtor in possession has the exclusive right, for 120 days, to propose a plan of reorganization. The effect of the exclusivity period and the *automatic stay* is to deny a creditor an *exit option* during the course of bargaining with the old shareholders. The exclusivity period is often extended, but termination of the exclusivity period itself does creditors little good if the bankruptcy judge is unwilling to confirm a plan of reorganization that allows the creditor to reach its *collateral.*

Exit option. The right of one player to terminate a bargaining game and receive some alternative *payoff.*

Expectation damages. Under the Anglo-American law of contracts, a person who breaks a contractual obligation must pay the innocent party an amount of damages sufficient to put that party in the same position the party would have been in had performance taken place as promised.

Extensive form game. An extensive form game contains the following elements: (1) the *players* in the game; (2) when each player can take an action; (3) what choices are available to a player when that player can act; (4) what each player knows about actions the other player has taken when that player decides what actions to take; and (5) the *payoffs* to each player that result from each possible combination of actions.

Externality. An externality exists whenever a person does not enjoy all the benefits or incur all the costs that result from the actions that person undertakes.

Farrell, Grossman, and Perry refinement. A refinement of the *perfect Bayesian equilibrium solution concept* that rules out *equilibria* where there is a deviation with a consistent interpretation. A deviation by one or more types has a consistent interpretation if exactly that type or types (and no others) of players strictly prefer the *payoff* from that deviation to their *equilibrium payoff* given that the uninformed party believes that this type or these types of informed players are among those who deviate in the same proportion that this type or these types bear to the population of informed players as a whole.

Fixture. See *real property*

Focal point. Also called a *Schelling point,* the combination of *strategies* that players are likely to choose because it is especially prominent under the conditions and culture in which the players find themselves.

Folk theorems. Folk theorems are central to understanding *infinitely repeated games.* The first of these theorems became common knowledge among game theorists before any paper was published using it. It and related theorems are called "folk" theorems just as well-known stories deeply embedded in a culture and whose authors are unknown are called "folk stories." The original folk theorem is that, when a *discount factor* is sufficiently high, there exists at least one *Nash equilibrium* that supports any combination of *payoffs* that is both feasible and gives each player at least as much as that player would receive if all the other players tried to minimize that player's *payoff*. Other folk theorems extend the idea to *subgame perfect Nash equilibria* and other *refinements*. Folk theorems show that a broad range of *strategy combinations* form an *equilibrium* to a *repeated game*. Hence, one cannot point to *tit-for-tat* or any other *strategy* combination as the combination of *strategies* the players are likely to adopt.

Free-riding. A player who is able to enjoy the benefit of an action that another player takes is a free-rider. Free-riding problems typically arise when there is a *public good*.

Game. See *extensive form game* and *normal form game*

Game theory. A set of formal tools for modeling the behavior of individuals whose actions are strategically linked. The *normal form game* and the *extensive form game* are the two basic kinds of game-theoretic models. When reduced to such basic elements as players, *strategies*, and *payoffs*, game-theoretic models resemble parlor games. It bears the name "game theory" for this reason.

Going-concern value. A firm may be worth more if the assets are kept in their current configuration under existing management than if sold in a liquidation. This amount is known as a firm's going-concern value. The difference between this value and the liquidation value is known as the going-concern surplus.

Grim strategy. Also known as a *trigger strategy*. The *strategy* in an *infinitely repeated game* in which a player adopts the *strategy* of cooperating in the first period and, in all subsequent periods, defecting if the other player has ever defected in any previous period, but otherwise cooperating. If the *discount factor* is sufficiently close to 1, the *strategy* combination in which both players adopt the grim strategy is one of infinitely many *subgame perfect Nash equilibria*.

Groves mechanism. See *Clark-Groves mechanism*

Imperfect information. See *perfect information*

Incomplete contracts. A contract that fails to realize fully the potential gains from trade in all states of the world. Such contracts are "insufficiently state contingent" in the sense that the terms of the contract are not optimal given all the possible states of the world. Legal scholars, however, sometimes use the idea of an incomplete contract to refer to a contract in which obligations are not fully specified. From this perspective, what matters is not whether the contract generates the optimal outcome in every possible situation, but whether it specifies the obligations of the parties in every possible situation.

Incomplete information. See *complete information*

Infinitely repeated game. See *repeated game*

Information set. A player in an *extensive form game* sometimes must move without knowing the course of the game up to that point. A player, however, knows the possible *nodes* to which the game could have led up to that point. These *nodes*, between which the player cannot distinguish, form an information set. When a player knows perfectly the course of the game up to a particular point, the information set of that player consists of a single *node*.

Intuitive criterion. See *equilibrium dominance*

Iterated dominance. A *solution concept* that takes the idea that players adopt *dominant strategies* and reject *dominated strategies* one step further. Not only will a player probably adopt a *strictly dominant strategy*, but a player will predict that the other player is likely to adopt such a *strategy* and will act accordingly. A player believes that other players will avoid *strictly dominated strategies*. Moreover, a player believes that the other player similarly believes that the first player will not play *strictly dominated strategies*. A player also thinks that the other believes that the first player thinks that the other player will not play *strictly dominated strategies*, and so forth ad infinitum.

Market preemption. See *preemption*

Matching pennies game. A two-by-two *normal form game* in which each player chooses heads or tails. The first player wins both pennies if both choose heads or both choose tails. The second player wins both pennies if one choose heads and the other tails. This game is one in which there is no *Nash equilibrium* which is a *pure strategy equilibrium*. Given any combination of *strategies*, one player is always better off changing to another. The only *Nash equilibrium* is a *mixed strategy equilibrium*.

Mechanism design. See *direct mechanism*

Mitigation. Under the law of contracts, the innocent party is obliged to take reasonable steps to minimize the damages that flow from the breach of contract.

Mixed strategy equilibrium. See *pure strategy equilibrium*

Moral hazard. A problem of hidden action. When the conduct of a particular player is not visible either to other players or to a court, a contract cannot be written contingent on whether that action has been chosen. Hence, a player will choose the course of action that is in the player's self-interest, and the other player will adjust the price of the contract accordingly. A person who acquires insurance, for example, is more likely to act carelessly, provided that the insurer's obligations under the contract cannot be conditioned on whether the insured exercises care.

Mortgage. See *security interest*

Nash equilibrium. The central *solution concept* in *game theory*. It is based on the principle that the combination of *strategies* that players are likely to choose is one in which no player could do better by choosing a different *strategy* given the ones the others choose. A pair of *strategies* will form a Nash equilibrium if each *strategy* is one that cannot be improved upon given the other *strategy*. We establish whether a particular *strategy* combination forms a Nash equilibrium by asking if either player has an incentive to deviate from it.

Call the *strategy* for the first player x^* and the *strategy* for the second player y^*. Now we need to ask whether, given that the second player will play y^*, the first player can do strictly better by switching to some *strategy* other than x^*. Similarly, we need to ask whether, given that the first player will play x^*, the second player can do strictly better by switching to some *strategy* other than y^*. If x^* is the best *strategy* for the first player in response to the second player's y^*, and y^* is the second player's best response to x^*, then this pair is a Nash equilibrium for the game. A large class of important games has at least one Nash equilibrium. If a proposed solution is not a Nash equilibrium, one player or another will have an incentive to deviate and adopt a different *strategy*. If the players are rational, they should choose *strategies* that are the best responses to what they predict the other will do.

Nash program. See *cooperative bargaining*

Nature. A hypothetical player introduced into an *extensive form game* to convert a game of *incomplete information* to one of *imperfect information*. Nature decides on the type of the player with *private information*. At every move, the uninformed player has an *information set* that contains several *nodes* (a different node for each type of player). Each *node* in the *information set* corresponds to each type of player that Nature might have chosen.

Negligence. The traditional common law test for tort liability. An injurer is liable for harm caused to a third party if the injurer was negligent—that is, if the injurer failed to act as a reasonable person would under the circumstances. The test for negligence adopted by Learned Hand in United States v. Carrol Towing Co., 159 F.2d 169 (2d Cir. 1947), is widely used in the law and economics literature. An injurer acts negligently when the marginal benefit that the injurer derives from a discrete action is less than the marginal harm that the action causes to someone else, adjusted by its probability.

Network externality. An externality that arises when the value that a user derives from a good increases with the number of others who also use it.

New value exception. Under the law as it existed before 1978, the old equityholders of an insolvent firm could, in a bankruptcy reorganization, obtain an equity interest in the reorganized firm by contributing cash or other "new value" that was tantamount to cash. The contribution of new value had to be equal to the value of the equity interest they received, and the shareholders could exercise this right only when the cash contributed was "substantial" and when it was "necessary." It is called the new value exception because the law of corporate reorganizations ordinarily adheres to an *absolute priority rule*, under which investors are entitled to be paid in full in order of their seniority before anyone junior to them receives anything. Whether the new value exception continues under current law remains unclear.

Node. The fundamental building block of the *extensive form game*. It is a point at which a player takes an action or the game ends. Each node is therefore either a *decision node*, a point at which a player must choose between different courses of action, or a *terminal node*, which sets out the *payoffs* that each player receives.

Noncooperative bargaining. A noncooperative bargaining model looks at the structure of the rules in a bargaining environment and derives the likely solution by examining how rational parties would best advance their self-interest in such an environment.

Nonverifiable information. Information is nonverifiable when an informed party lacks any direct means of conveying the information to others. In such cases, uninformed players and third parties, such as courts, can only infer the information from the actions of the informed player.

Normal form game. Sometimes also called the *strategic form* game, this game consists of three elements: (1) the players in the game; (2) the *strategies* available to the players; and (3) the *payoff* received by each player for each possible combination of *strategies*.

Observable information. Information is observable, but not *verifiable*, when other players can identify another player's type, but a third party, such as a court, cannot.

Off the equilibrium path. See *equilibrium path*

Optimism model. This optimism model of litigation assumes that parties fail to settle because one or both of the litigants has views about the likelihood of success that are too optimistic.

Pareto-optimal. A *solution* to a game is Pareto-optimal if there is no other combination of *strategies* in which one of the players is better off and the other players are no worse off.

Payoff. The utility a player derives under a particular combination of *strategies*. Because we make the assumption that parties are *risk neutral* when such an assumption does not alter the character of the results, we are ordinarily able to express payoffs in arbitrary dollar amounts. What matters is not the amount of a payoff a player receives, but rather the amount of that payoff relative to what the player would receive from choosing other *strategies*.

Perfect Bayesian equilibrium. Under this *solution concept*, every player begins the game with *beliefs*. These *beliefs* must be updated in light of *Bayes's rule*, and the actions that a player takes in equilibrium must be *sequentially rational*. They must be optimal given the *beliefs* of the player and the actions of all the other players.

Perfect information. Information is perfect when each player knows the full history of the game up to that point. Information is *complete* but *imperfect* when each player knows the *strategies* available to each player and the *payoffs* to every player for every combination of *strategies*, but at least one player must make a move without knowing the complete history of the game up to that point. A *prisoner's dilemma* or a *stag hunt* is an example of a game of *complete* but *imperfect information*. Each player must move without knowing what the other player has done.

Perfect tender rule. The right a buyer enjoys in a contract for the sale of goods to reject the goods and return them to the seller if they are nonconforming or if they fall short of being the goods that the seller promised in the contract.

Personal property. Property that is not *real property* or a *fixture*.

Player. Each individual who interacts strategically with another is represented in *game theory* as a player in a *game* who must choose among different *strategies*.

Pooling equilibrium. A *solution* to a game in which players of different types adopt the same *strategies* and thereby prevent an uninformed player from drawing any inferences about an informed player's type from that player's actions.

Predatory pricing. A practice in which a firm charges low prices for a time in order to drive out competing firms so that it can raise prices to even higher levels after the competing firms leave. A firm will typically be found to have engaged in predatory pricing only if it can be shown that it set prices below its own short-run average variable cost and can reasonably expect to recoup losses incurred while predating after competitors have left the market.

Preemption. A firm with market power preempts another when it expands its capacity before it otherwise would as a way of deterring others from entering the market.

Prisoner's dilemma game. A paradigmatic, two-person, two-strategy *normal form game* of *complete* but *imperfect information*. In this game, the *strategy* combination that is in the joint interests of the players (to remain silent in response to a prosecutor's questions) is not played, because each player finds that the *strategy* of remaining silent is *strictly dominated* by the other *strategy* (confessing). This game is emblematic of some collective action problems in the law in which individual self-interest leads to actions that are not in the interest of the group as a whole.

Private information. Information possessed by fewer than all of the players in a game. Private information may be *verifiable* or *nonverifiable*.

Public good. A good is a public good when the marginal cost of supplying it to an additional consumer is close to zero. A radio broadcast is a prototypical public good. Although the program is costly to produce, once a signal is broadcast, it costs nothing for each additional viewer to tune in to a program.

Pure strategy equilibrium. A *Nash equilibrium* in which each player adopts a particular *strategy* with certainty. In a *mixed strategy equilibrium,* one or more of the players adopts a *strategy* that randomizes among a number of pure *strategies.*

Real property. Real property consists of land and of *personal property* (such as bricks and mortar) that have become incorporated into the realty. *Fixtures* lie between real property and *personal property.* They are *personal property* (such as elevators) permanently attached to the realty but capable in principle of being separated from it.

Refinement. *Solution concepts* frequently lead to multiple *equilibria.* Refinements are modifications of *solution concepts* that eliminate *equilibria* that players are unlikely to adopt. An *equilibrium* may be unlikely because it rests on a threat that is not *credible* or an implausible *belief. Subgame perfection* and *equilibrium dominance* are two important refinements.

Reliance damages. A party who is the victim of a breach of contract receives reliance damages if the damages are sufficient to put the party back in the same position the party would have been in had the contract never been entered into. Reliance damages reimburse the innocent party for the costs incurred as a result of entering into the contract. Anglo-American law typically awards *expectation damages* rather than reliance damages. In many contexts, however, the two are the same.

Renegotiation-proof equilibrium. An *equilibrium* of an *infinitely repeated game,* such as the *prisoner's dilemma,* that ensures that the nondeviating party still has the incentive to carry out the punishment in the event that it is called upon to mete it out. When the nondeviating party punishes the other party for deviating, it must receive at least as much as it would receive if both parties played the cooperative *strategy.*

Repeated game. A game in which players play what could be a stand-alone game multiple times in succession. An *infinitely repeated game* is a repeated game with no terminus. A game with an indefinite terminus can be modeled as an *infinitely repeated game.*

Revelation principle. A principle which provides that, for any allocation mechanism that leads to a particular *equilibrium,* there also exists a *direct mechanism* in which the players truthfully reveal their type that implements the same *equilibrium.* If no *direct mechanism* in which parties truthfully reveal their type exists that reaches a particular *equilibrium,* no other scheme that we can devise can reach it either. The revelation principle is useful in modeling public goods, auctions, regulation, and any other environment in which the problem is one of designing an allocation mechanism in the presence of private, *nonverifiable information.*

Risk neutrality. A preference with respect to risk such that a person is indifferent between a fixed sum and any lottery with the same expected payoff. Even though individuals are typically risk averse and prefer fixed amounts to uncertain payoffs, the assumption of risk neutrality frequently does not alter the character of an economic model.

Rubinstein bargaining game. A two-person, alternating offer bargaining game with an infinite horizon. The game might take the following form: Two players are seated at a table. In the center of the table is a dollar bill. The players must negotiate with each other and agree on a way of dividing the dollar. One makes an offer that the other can accept or reject. Once a player rejects an offer, that player can make a counteroffer that the first can in turn accept or reject. When they reach some agreement on how to divide the dollar, they will each receive their respective share of the dollar. Unless and until they reach agreement, however, they receive nothing. Delay matters because a dollar today is worth more than a dollar at some later time.

Schelling point. See *focal point*

Screening. Actions that uninformed players can choose that lead informed players to act in a way that allows the uninformed player to infer the informed party's type. For example, an insurance company may use a high deductible to screen out different types of drivers. Those least likely to have an accident may be most likely to pay for a policy with a high deductible.

Security interest. A lender who extends credit may bargain for the right to claim and seize particular property in the event of default. Such a right is called a security interest. The property securing the debt is called *collateral.* In the context of a corporate debtor, the principal effect of a security interest is to give one creditor the right to be paid (up to the amount of its loan) before another creditor. A security interest when the *collateral* is *real property* is also called a *mortgage.*

Separating equilibrium. A *solution* to a game in which players of different types adopt different *strategies* and thereby allow an uninformed player to draw inferences about an informed player's type from that player's actions.

Sequential rationality. Part of the *perfect Bayesian equilibrium solution concept.* A player's behavior is sequentially rational if the actions that the player takes are optimal given the *beliefs* of the player and the actions of the other players.

Signaling. Strategy choices by those who possess *nonverifiable information* that convey information.

Solution concepts. The means by which games are solved. Solution concepts are general precepts about how rational parties are likely to choose *strategies* and about the characteristics of the *strategies* that the players are likely to adopt given their goals.

Specific performance. Ordinarily under Anglo-American contract law, a party is entitled only to *expectation damages* for breach of contract. Under some circumstances, however, a party may be entitled to a court order forcing the other party to perform as promised. A right to such an order is called a right to specific performance.

Stag hunt game. A paradigmatic, two-person, two-strategy *normal form game* of *complete* but *imperfect information.* A stag hunt is a coordination game with multiple *Nash equilibria.* The *strategy* that each player adopts is good or bad depending on what the other does. The game involves two hunters. Each prefers hunting stag to hunting hare, but only if the other hunts stag as well. Because it is also a *Nash equilibrium* for both to hunt hare, neither can be certain that the other will hunt stag.

Stationary game. An infinite-horizon, repeated game is stationary if the continuation game at some subsequent *node* is identical to the game at the initial *node.* A *Rubinstein bargaining game* is one such game. Because both parties are fully informed at the outset, neither party learns anything over the course of negotiations. Moreover, the *discount factor* throughout the game is constant. When the players begin a new round of bargaining, they are worse off than they would have been if they had reached a deal at the initial *node,* but the game that they play looks exactly the same. There is still a dollar in the middle of the table, and they know no more or no less than when they started.

Statute of Frauds. Many contracts under Anglo-American law, including contracts for the sale of goods over $500, must be evidenced by a writing signed by the party against whom enforcement is sought. The original seventeenth-century English statute requiring a writing was called the Statute of Frauds, and all similar rules now bear this name.

Strategic form game. See *normal form game*

Strategy. A strategy for a player specifies the actions that a player takes at every *information set* in a game, even when it is an *information set* that cannot be reached because of previous actions the player has taken.

Strategy space. The set of all the *strategies* available to a player.

Strict liability. Under a regime of strict liability, an injurer is liable for the harm that the injurer's actions cause to another, regardless of how carefully the injurer acted.

Subgame. A *node* or a set of *nodes* of an *extensive form game* that can be viewed in isolation. More formally, a subgame of a game in the *extensive form* is any part of a game that meets the following three conditions: (1) it begins at a *decision node* that is in an *information set* by itself; (2) it includes all the *decision nodes* and *terminal nodes* that follow that *decision node* in the game and no others; and (3) no *nodes* belong to an *information set* that includes *nodes* that do not also follow the *decision node* that begins the subgame.

Subgame perfect Nash equilibrium. A *Nash equilibrium* is subgame perfect if the players' *strategies* constitute a *Nash equilibrium* in every *subgame*. In other words, one imagines that every *subgame* exists by itself and asks whether, when this *subgame* is examined in isolation, a proposed *equilibrium* is one in which both players have adopted *strategies* that are optimal given the *strategy* of the other.

Substantial performance. A person who contracts for the construction of a building has a right to insist on only substantial performance. If a builder fails to perform as promised, but nevertheless performs substantially, the buyer's remedy is to sue for damages. The buyer does not enjoy the right to withhold payment until the builder corrects the problem and completes performance in accordance with the contract.

Supergame. A game that consists of an infinite number of repetitions of another game.

Terminal node. See *node*

Tit-for-tat. The *strategy* in an *infinitely repeated game* in which a player embraces the *strategy* of cooperating in the first period and, in all subsequent periods, defecting if the other player defected in the immediately preceding period, but otherwise cooperating. A player who defects is punished in the next period. Unlike with the *grim strategy*, however, the player who punishes will return to the cooperative equilibrium if the other player once again cooperates. If the *discount factor* is sufficiently close to 1, the tit-for-tat strategy is one of infinitely many *Nash equilibria* in these repeated games. Experimental work suggests that tit-for-tat is a successful *strategy*. There is, however, no reason in principle to predict the *strategy* combination in which both players adopt tit-for-tat as the *equilibrium* of these games, given the basic *folk theorems*. Moreover, tit-for-tat does poorly in environments in which information is imperfect.

Tragedy of the commons. The collective action problem that arises when individual shepherds graze sheep on a common pasture. Individual shepherds do not bear the full costs of adding an additional sheep to their flocks. Because each additional sheep benefits one shepherd, but makes grazing conditions worse for all the shepherds as a group, the self-interest of individual shepherds leads to the overgrazing of the pasture.

Tremble. A deviation from an *equilibrium* that results when a player makes a mistake. (One can think of a player's hand "trembling" while making a move.) When we hypothesize that play *off the equilibrium path* is the result of a mistake, we can examine the player's possible courses of action without taking into account how the game reached this point. (If the other player made an out-of-equilibrium move deliberately, this fact would have to be taken into account in deciding what action to take.)

Trigger strategy. See *grim strategy*

Unraveling result. When information is *verifiable* and the uninformed party knows that the other party possesses information, the informed party is likely to disclose the information. The informed parties with the most favorable information disclose the information to prevent being pooled with the rest. Those remaining with the most favorable information will then disclose. The process continues until everyone discloses except for the person with the worst information. At this point, the other player can draw an inference about that player's type from silence.

Verifiable information. Information is verifiable if the party who possesses the information is able to convey that information readily to another player and to a third party, such as a court.

von Neumann-Morgenstern expected utility theory. Von Neumann and Morgenstern proved that, when individuals make choices under uncertainty in a way that meets a few plausible consistency conditions, one can always assign a utility function to outcomes so that the decisions people make are the ones they would make if they were maximizing expected utility. This theory justifies our assumption throughout the text that we can establish *payoffs* for all *strategy* combinations, even when they are mixed, and that individuals will choose a *strategy* based on whether it will lead to the highest expected *payoff*.

Zero-sum game. A game in which the increase in the *payoff* to one player from one combination of *strategies* being played relative to another is associated with a corresponding decrease in the *payoff* to the other. If we sum the *payoffs* that all the players receive under each combination of *strategies*, we find that this sum is the same. One player does better only if another player does worse. It is called a zero-sum game because one can renormalize the *payoffs* in every cell and ensure that the sum of the *payoffs* to the players is always 0 under every *strategy* combination without changing the character of the game.

Index